MW01493087

Zillions of Practice Problems

Pre- Algebra 2 with Economics

Stanley F. Schmidt, Ph.D.

Polka Dot Publishing

ISBN: 978-1-937032-60-9

Printed and bound in the United States of America

Polka Dot Publishing   Reno, Nevada

To order copies of books in the Life of Fred series,

visit our website PolkaDotPublishing.com

Questions or comments? Email the author at lifeoffred@yahoo.com

First printing

Zillions of Practice Problems Pre-Algebra 2 with Economics was illustrated by the author with additional clip art furnished under license from Nova Development Corporation, which holds the copyright to that art.

for Goodness' sake

or as J.S. Bach—who was
never noted for his plain
English—often expressed it:

Ad Majorem Dei Gloriam
(to the greater glory of God)

If you happen to spot an error that the author, the publisher, and the printer missed, please let us know with an email to: lifeoffred@yahoo.com

As a reward, we'll email back to you a list of all the corrections that readers have reported.

How This Book Is Organized

Life of Fred: Pre-Algebra 2 with Economics has 34 chapters before the final bridge. So does this book.

As you work through each chapter in *Life of Fred: Pre-Algebra 2 with Economics* you can do the problems in the corresponding chapter in this book.

Each chapter in this book is divided into two parts.

✶ The first part takes each topic and offers a zillion problems.

✶ The second part is called the Mixed Bag. It consists of a variety of problems from the chapter and review problems from the beginning of the book up to that point.

Please write down your answers before turning to the back of the book to look at my answers. If you just read the questions and then read my answers you will learn very little. As my mother used to tell me,

*Everything comes to those
who wait,
as long as you are
working like crazy
while you wait.*

Chapter One
Summertime

First part: Problems from this chapter

104. Definition: A **function** is any rule that associates each element of the first set (the domain) with exactly one element of the second set (the codomain).

Suppose the first set (the domain) is the set of all teachers at KITTENS University. Suppose the second set (the codomain) is the set of all numbers.

Is this a function? *Associate to each teacher the salary that they are now making each month.*

235. Let's switch things around. Suppose the domain is the set of all numbers and the codomain is the set of all teachers at KITTENS.

Is this a function? *Associate to each number the teacher that is now making that salary each month.*

314. Suppose that the first set (the domain) is the set of students at KITTENS University. Suppose the second set is the set all humans. Why is this not a function: *Assign to each student their favorite ice hockey player.*

404. Again, let the domain be the set of students at KITTENS and the codomain be the set of all humans. Is this a function? *Associate to each student their biological mother.*

546. (continuing the previous problem) Mike and Ike are students at KITTENS. They are twins. Make an argument why *Associate to each student their biological mother* is still a function.

❦❦ *Functions are tough. Let's take a break and do a little arithmetic.* ❦❦

622. $\frac{7}{8}$ is how much larger than $\frac{3}{5}$?

772. If it takes 38 drops of sweat to learn how to do percents, and it takes 45% more sweat to learn about functions, how many drops will that be?

Chapter Two

Getting Paid in Nickels

First part: Problems from this chapter

144. Lucy always loved trains.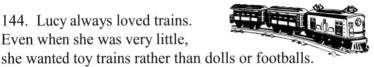
Even when she was very little,
she wanted toy trains rather than dolls or footballs.

 For Lucy's first birthday her parents gave her four dolls.

"Yucky!" she exclaimed. That was the first word she ever spoke. Up to that moment she had been silent. She had been thinking about trains and had never needed to tell anyone else about her thoughts.
She repeated, "Yucky!" Lucy's mother dutifully wrote in Lucy's baby book under Baby's First Word: *Yucky*.*

Lucy handed her 4 dolls to her brother and took his engine and two train cars in exchange. Luke, her brother didn't make a fuss. He knew that one-year-olds often went through phases. Luke was certain that Lucy was just going through a "train phase" and that she'd be interested in other things. Then he could get his train cars back. There was no need to force her to give back the train and make her cry. Instead, he just pulled 51 other train cars out of his closet and played with them.

Lucy's eyes lit up. She ran to her bedroom and gathered up some of her other yucky dolls. She was going to trade 4 dolls for every 3 train cars. How many dolls would she need?

300. There was one doll that Lucy would never give up. It was her engineer doll—the guy who drove the train. Lucy called him Ginny, not realizing that Ginny is a girl's name. For Lucy, the name Ginny was short for Engineer.

Ginny's head was 5 cm (cm = centimeter) tall. That was two-ninths of his total height. How tall was Ginny?

The conversion factor will be $\dfrac{5 \text{ cm}}{2/9 \text{ total height}}$

---

* Some people tell me, your author, that my first word was pizza.

Second part: the 𝔐ixed 𝔅ag: a variety of problems from this chapter and previous chapters

316. Lucy needed passengers for her long train. Her brother, Luke, had 72 dolls. She took them back and gave Luke nothing. When you are a one-year-old, you think that anything you can grab is yours. Thieves who are 17-years-old are just one-year-olds in disguise.

Lucy had an engine and 53 train cars and 73 dolls on her bedroom floor.

Ginny, the engineer
2 pig dolls—Porky and Ima
68 girl dolls
1 boy doll
1 crocodile doll named Scizzzors   (She spelled it with three z's.)

1 red engine
1 green train car
1 blue train car
1 brown train car
50 gray train cars

If the domain (the first set) is the dolls and the codomain is the train cars, is this a function? *Lucy put Ginny in the engine, the pig dolls in the brown car, the 68 girl dolls in the green car, and the boy doll in the blue car. (She left the 50 gray cars empty.)*

455. On Lucy's first birthday her parents had given her four dolls. They had also given her some pink nail polish. They hoped that these gifts would encourage her to do "girl stuff." Her mom showed her how to paint one of her nails.

Lucy thought What a waste of paint! I know what this should be used for. That is how one of Luke's train cars became pink. Luke did not like this at all.

The bottle was $\frac{9}{10}$ full when Lucy started painting the car. (That's because her mom had used $\frac{1}{10}$ of the bottle to paint one of Lucy's nails.) When Lucy was done, there was $\frac{1}{6}$ of the bottle left. How much had she used in painting the train car pink?

Chapter Three
Finding a Job

First part: Problems from this chapter

172. Lucy was almost potty trained. She only made "mistakes" a couple of times each day. But she had thoughts that many other one-year-olds don't have: Someday I want to own a railroad—a real railroad. It will run from New York all the way to San Francisco and then on to Hawaii.* People will happily trade their money for what the Lucy Railroad will offer them. Then I can buy a big house, a lot of friends, a doggie, and a husband.**

If you are going to get money from others (and are not going to steal it), you have to offer to trade what you have for their money. There are only two things you can offer.

Two Things You Can Offer

I. Your things (your dolls, your silverware, your stamp collection, your oil paintings, your dishes . . .). This is what a garage sale or a yard sale is all about.

II. Your _____ .
fill in one word

273. When people are going to trade their money for the Two Things You Can Offer (see the previous question), they are looking at two things: ① the price and ② the quality. (This is called the *second key point of economics*.)

If Lucy wanted to make money right now—when she is only 1 year old—she didn't have much to sell, so the only thing she could offer is her labor. (Again, see the previous question.) She thought of being a carpenter. Looking at the second key point of economics, why wouldn't that work?

---

* No one had told her about the Pacific Ocean.

** One-year-olds think that money can buy everything.

Second part: the 𝔐ixed 𝔅ag: a variety of problems from this chapter and previous chapters

329. Solve $9y - 13 = 3y + 17$

456. Solve $12x - 4x = 5$

511. Lucy knew that right now her main job was to get physically strong and to learn everything possible about railroading. Then when she grew up she could create the Lucy Railroad.

Kids do a lot of learning by playing. For kids, playing is not wasting time. When kids play with dolls, they are learning about being parents. When Lucy puts her crocodile Scizzzors doll in one of her cattle cars, she's learning how to transport livestock on a railroad. If she is charging 5¢ per foot to transport Scizzzors across her bedroom floor, how much would it cost to

move Scizzzors 3 meters? (Start with $\dfrac{5¢}{1\ \text{foot}}$

and your conversion factor will be based on 1 foot ÷ 0.3048 meters.)

649. Lucy thought for a moment: I could get 4¢ for moving Scizzzors three meters. How far would Lucy Railroad have to move that crocodile in order to make a million dollars?

777. Lucy got excited.   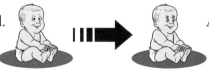   A little too excited.

She wet her diaper. Kids who learn about fractions before they are potty trained talk about wetting their denominator (since a denominator is the bottom of a fraction). Lucy thought about wetting her caboose (since a caboose is the rear car on a freight train).

Six diaper changes per day. Each one take 5 minutes. Her mom is awake 16 hours per day. What percent of her waking day is spent on Lucy's caboose?

Chapter Four
From English to Equations

First part: Problems from this chapter

120. Lucy had Luke's engine and 53 train cars. The engine, which had a motor in it, cost $3.39. The other non-engine cars each cost x dollars. Luke had bought the whole collection for $129. How much did each of the other cars cost?

> Show all the steps. Do not just jump to writing the equation.

227. Almost all of Lucy's 73 dolls each weighed 113 grams. The only exception was Scizzzors, the crocodile doll. He was made out of steel and weighed 864 grams. All 73 dolls together weighed 9,000 grams. How much did each of the 72 non-Scizzzors dolls weigh?

333. One pound is approximately equal to 454 grams. In the previous problem we found that one of the non-Scizzzors dolls weighed 113 grams. Does a quarter-pound hamburger weigh more or less than one of these dolls?

457. Scizzzors weighs 864 grams. A regular doll weighs 113 grams.

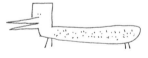

To the nearest percent, how much more does Scizzzors weigh than a regular doll? (The answer is not 7% or 765%.)

512. Lucy didn't like the way Scizzzors looked. She thought that crocodiles should have tails. She got a paperclip and attached it to the doll. That increased its weight by 5% Now she thought he looked much more like a real crocodile. How much did Scizzzors now weigh?

Second part: the 𝔐ixed 𝔅ag: a variety of problems from this chapter and previous chapters

700. Eleven ducks plus an $8 sack of duck food cost the same as five ducks and a $22.58 sack of duck food. How much does a duck cost?

     Let d = the cost of a duck.

702. Let T = the set of all train cars that Lucy is playing with. Let W = the whole numbers, which is {0, 1, 2, 3, 4, . . .}.

     Is this a function from T to W? *Assign to each train car the number of wheels that it has.*

836. (continuing the previous problem) Is this a function from T to W? *Assign to each car its weight in grams.*

902. Lucy took Scizzzors outside and put him in the garden. She thought Now he can get a little breath of fresh air and maybe eat some insects. That will be good for him.

     It was Tuesday. She left him there all day. Let the first set S be the set that contains just Scizzzors—{Scizzzors}—and the second set be the set of all numbers R.

     Is this a function from S to R? *Assign to each member of S, its temperature on that Tuesday.*

     (In this problem to be definite, we'll measure the temperature using the thermometer that Lucy's mom has.)

taking Scizzzors' temperature

First part: Problems from this chapter

161. Lucy had her second birthday. She's growing up. She was potty trained and no longer needed to get her caboose changed in the switching yard. She did a lot of thinking about trains, and she now could speak three words: **yucky**, **locomotive**, and **caboose**. Locomotive is the fancy word for a train's engine.

What percent increase was it in going from {yucky} to {yucky, locomotive, caboose}?

327. Lucy had gone beyond just reading magazine articles about trains. There was very little on television about trains. She saw one kids' program about trains, and they talked about choo-choo trains. That felt so juvenile to her.

For her second birthday, she wanted **books** about trains. Having a three-word spoken vocabulary, it was impossible for her to tell that to her parents.

She got six more dolls for her birthday. For Lucy they were new passengers on Lucy's railroad.

She got a tube of lipstick. They were hoping that she would put it on her mouth, and that would encourage her to speak more. She looked at that lipstick and said, "Yucky."

Lucy's goal in life was to create the Lucy Railroad. The definition of *tool* is something that helps you accomplish your goal more easily. In a sentence or two explain why learning lots about trains is a tool for Lucy.

334. We want education to be **fast**, cheap, and pleasant.

Let's start with pleasant. Which is these is most pleasant?

    A) Reading
    B) Watching movies, television, and videos
    C) Going to college to hear lectures

468. Education is a great tool. We want it to be *fast*, cheap, and pleasant. Which of these is *not* cheap?

    A) Reading books
    B) Watching movies, television, and videos
    C) Going to college to hear lectures

514. Education is a great tool. We want it to be *fast*, cheap, and pleasant. Which of these is *fast*?

    A) Reading books
    B) Watching movies, television, and videos
    C) Going to college to hear lectures

623. Lucy was going to get rid of her old train magazines. She had read them all many times and was now ready for a real education: books.

    13 issues of *Locomotive Monthly* and 31¢ were worth the same as 4 issues and $13. How much was each issue worth?

    Start your solution by writing "*Let x = the price of one issue.*"

Second part: the 𝔐ixed 𝔅ag: a variety of problems from this chapter and previous chapters

775. On the day after Lucy's second birthday she learned her fourth word: *Toot!* Is *fourth* an ordinal or a cardinal number?

831. She ran around the house yelling *Toot! Toot! Toot! Toot! Toot! Toot! Toot! Toot! Toot! Toot! Toot! Toot! Toot! Toot! Toot! Toot!*

This drove her brother Luke nuts. He thought to himself I wonder if my two-year-old sister is dumb. She's two years old and she only can say four words. Maybe they will let her in kindergarten when she is eight years old. She'll graduate from high school at the age of 40.

Meanwhile Lucy sat down with Prof. Eldwood's book, *Elementary Guide to Railroad Management,* and read: It is essential in the selection of a locomotive—whether steam or diesel—for use in sorting yards or coach yards that attention be paid to the horsepower requirements as given by the formula: HP = 377n where n is the maximum number of cars to be pulled.

If Luke knew what Lucy was reading, he would have a much different opinion of her mental abilities.

What is the horsepower requirements of a locomotive that will be pulling 4 cars?

904. The set S of Lucy's speaking vocabulary is {yucky, locomotive, caboose, toot}. What is the cardinality of this set?

931. $3\frac{3}{4} + 4\frac{7}{8}$

Chapter Six
Which Job?

First part: Problems from this chapter

118. Prof. Eldwood has written a lot of books, but his writing skills are poor. In his *Elementary Guide to Railroad Management* he wrote: In the event it happens that two or more trains are situated on the same track and heading towards each other at a high rate of speed, it should be noted that it is essential that in the interests of good railroad management that steps be taken to avoid the situation in which impaction occurs.

 No wonder people don't like to read his books! Clean up that sentence as much as you can.

251. Lucy found six books on trains at the KITTENS library and stuffed them into her 0.58-pound backpack. The books all weighed the same. She handed the 10 pounds to her brother to carry. How much did each book weigh?

391. Luke took Lucy's 10-pound backpack and put his 32 comic books in it. The whole thing now weighed 22 pounds. The comic books all weighed the same. How much did each comic book weigh? Do this problem using fractions instead of decimals.

$+ \ 10 \ = \ 22$

401. (continuing the previous problem) Convert the weight of a comic book into ounces. (1 pound = 16 ounces)

543. Luke had 9 pennies and a 4.6-gram ball of gum in his left pocket. That weighed the same as the 6 pennies and the 16-gram rock he had in his right pocket.

 How much does a penny weigh?

Second part: the 𝔐ixed 𝔅ag: a variety of problems from this chapter and previous chapters

625. One copy of an Adventure Duck comic book costs 36¢.

When Luke bought 32 copies, they gave him a 16⅔% discount. (You remember that 16⅔% is the same as $\frac{1}{6}$ don't you?)

How much did he pay for a each comic book? Work in fractions rather than decimals.

780. Let D = the set of all 32 Adventure Duck comic books that Luke bought. Let T = the set of train books that Lucy got from the KITTENS library.

In case you are interested, they were:

Prof. Eldwood's *Trains & Today*
Prof. Eldwood's *Choo-choo Trains—A Sociological Perspective*
Prof. Eldwood's *Locomotive Electrical Diagrams*
Prof. Eldwood's *Your Future in Railroading*
Prof. Eldwood's *Why Railroad and Rich Both Start with R*
Prof. Eldwood's *Dining Recipes for Railroad Use*

Is this a function? *Assign to each member of D to Prof. Eldwood's* Locomotive Electrical Diagrams.*

907. Now let's turn your brain around. Suppose T (as defined in the previous problem) is the domain and D is the codomain. Suppose the first comic book in D is very special to Luke. He has read it so many times that only it has a torn cover. Is this a function? *Assign Prof. Eldwood's* Locomotive Electrical Diagrams *to the first comic book in D* (the one with the torn cover).

---

* Normally, book titles are *italicized*. (When you are using a typewriter, you <u>underline</u> instead of italicizing.)

However, when the sentence is already italicized, the book title is then not italicized. English has a lot of rules. Math has one: Don't divide by zero.

Chapter Seven
Fred's Career as a Jockey

First part: Problems from this chapter

160. Let D = the set of all 32 Adventure Duck comic books that Luke bought. Let T = the set of 6 train books that Lucy got from the KITTENS library.

　　　We know that assigning every member of D to Prof. Eldwood's *Locomotive Electrical Diagrams* book is a function.

　　　Your question: Is this function one-to-one?

217. (continuing the previous problem) Is there a way to create a 1-1 function from D to T?

315. For many students is the concept of function one of the hardest topics they have met in mathematics? □ yes   □ yes

　　　Not every question about functions is hard. Very few readers got #315 wrong.

462. Let C = {purple, yellow}. Let J = the set of the 31 jockey hats that Fred has.

　　　Is it possible to create a function from C to J that is *not* 1-1?

548. Is this a function from {7, 8} to {#, @, 5}?
$$7 \rightarrow \#$$
$$8 \rightarrow \#$$

626. Is this a function?

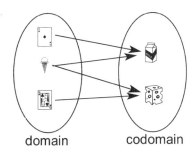

domain　　　codomain

Second part: the 𝔐ixed 𝔅ag: a variety of problems from this chapter and previous chapters

701. Create a function in which the domain is {5, 7} and the codomain is {egg, top}.

782. Lucy had dreams of creating Lucy Railroad. It would be bigger than any other railroad in the country.

    If the second biggest railroad had 6,308 cars, she knew she would have 250% more cars.

    How many cars would Lucy Railroad have?
(The answer is not 15,770.)

832. Solve $700x + 398 = 936x + 67.6$

906.

diagram #25506

Lucy was carefully studying Prof. Eldwood's *Locomotive Electrical Diagrams* book. She wanted to know as much as possible about all the facets of railroading.

    Lucy explained diagram #25506 to Scizzzors.* When you have smart dolls, they can learn it as fast as you can teach it.

    It took Lucy 8 minutes to understand this diagram. Scizzzors looked at it with his steel eyes and understood it in 5 seconds.

    She knew that when she created the Lucy Railroad she would hire Scizzzors to work in the electrical department.

Eight minutes is 480 seconds. $(8 \times 60 = 480)$
480 is what percent more than 5?

---

* The best way to learn something is to teach it to others. School teachers often know more about the subject matter than their students only because they have taught it.

    If you really want to learn *Life of Fred* material, teach it to your younger brothers or sisters—or teach it to your parents.

Chapter Eight
In Business

First part: Problems from this chapter

166. On the Internet Lucy saw the bargain that she couldn't live without. She knew that Lucy Railroad would require train stations, and train stations would require statues.

Train Station Statues! On sale now! $500 each!

Statues add class to anything she thought. I'm going to have lots of statues in my house when I grow up.

Lucy sent in $10,000. (She put it on her credit card since she didn't have a checking account yet.) There was a $1,000 shipping charge.

How many statues would she get?

229. Actually, it wasn't Lucy's credit card. It was her parents' card.[*] Several weeks later at the breakfast table:

Mom: (to Dad) Honey. Did you look at our credit card bill?

Dad: (while eating his eggs and sausages) What about it?

Mom: There's this charge for $10,000 from the Train Station Statuaries Company. Did you buy something?

Dad: (He shook his head, because his mouth was full.)

Both of them turned their eyes toward Lucy who was eating her Toot!® brand cereal.

If the domain is the two parents and the codomain are Lucy and Luke, would this be a function: *Assign each parent to Lucy.*

330. Toot!® brand cereal is the cereal for people who love railroads. Each

little bite of cereal is a little locomotive that is 95% oats and 5% engine oil. Their slogan: It tastes like a real locomotive! A handful of Toot!® (130 grams) would contain how much engine oil?

---

* If only one parent had owned the card, then it would be her parent's card.

Second part: the 𝔐ixed 𝔅ag: a variety of problems from this chapter and previous chapters

488. Two-year-olds like to play with their food. She didn't realize that both of her parents were staring at her with a bit of anger. This wasn't the first time that their "train kid" had made a purchase on their credit card.

Lucy was taking each  and with a black felt-tip pen was changing it to

She could redecorate 12 locomotive every 25 seconds. How long would it take her to redecorate all 126 locomotives in her cereal bowl? (Use a conversion factor.)

513. After a little more than four minutes Lucy had finished renaming the locomotives. Her mom said, "𝗟𝗼𝘂𝗰𝘆." With that tone of voice everyone at the table knew that Lucy was in trouble. "Normal kid" Luke excused himself from the table. He didn't want to hear what was coming next to the "train kid." (What follows is an imaginary conversation since Lucy doesn't yet have a large speaking vocabulary.)

Mom: 𝗟𝗼𝘂𝗰𝘆, 𝗱𝗶𝗱 𝘆𝗼𝘂 𝘀𝘁𝗲𝗮𝗹 𝗼𝘂𝗿 𝗰𝗿𝗲𝗱𝗶𝘁 𝗰𝗮𝗿𝗱?

Lucy: No, Mom. You told me not to do that ever again after I bought the iron railing last month for my train stations.

Mom: 𝗧𝗵𝗲𝗻 𝗵𝗼𝘄 𝗱𝗶𝗱 𝘆𝗼𝘂 𝗺𝗮𝗸𝗲 𝘁𝗵𝗶𝘀 𝗽𝘂𝗿𝗰𝗵𝗮𝘀𝗲? She held the credit card statement up to Lucy's face.

Lucy smiled and explained that she had memorized the 16 digits on the credit card so that she would never have to steal the card again. From her two-year-old point of view, she had obeyed her mom. She hadn't stolen her mom's credit card.

Two-year-olds can be very literal-minded.

Lucy's allowance was $1.25 per week. How long would it take for her to pay off the $10,000 purchase?

652.   Dad: Lucy, you bought $10,000 from the Train Station Statuaries Company. What did you buy?

Lucy: That's silly, Daddy. I bought train statues. I didn't buy pizzas from a train station statuaries company. [Lucy giggled.]

Dad: For $10,000 you must have bought a million of those toy statues.

Lucy: I only got 18 of them.

Dad: Eighteen! They must be made out of gold. Where are they? I want to get them returned. [Dad starts to head toward Lucy's bedroom.]

Lucy: They are made out of marble. They are not in my bedroom, because they wouldn't fit.

She pointed to the credit card statement that her mother was holding and that told where she had put the 18 statues and the iron railing.

The fifth line on the credit card statement read:

Sammy's Stuff-It-In Storage Units,
monthly rental
fee.......................................$125.

eight feet tall

If you have a $1.25/week allowance, how long could you make payments on a $125/month storage unit?

770. Her dad asked, "Do you have anything else in that storage unit?"

Lucy replied, "No. That's all I have in that unit. The other stuff wouldn't fit in *that* unit."

Both parents fainted. Luke dropped the comic book that he was reading as he realized what Lucy meant. He looked at the credit card statement that lay on the floor.

Sammy's Stuff-It-In Storage Units,
monthly rental fee.......................................$125.
monthly rental fee.......................................$125.
monthly rental fee.......................................$125.
monthly rental fee.......................................$125.

Chapter Eight   In Business

His sister had rented four storage units. He did a quick calculation. My sister is 2. In 16 years she'll be on her own. Sixteen years of storage rental at $500/month (125 × 4) will cost a total of . . .

Finish this calculation for him using conversion factors.

Luke was feeling like he had a very expen$ive sister, and that his parents would be so broke that his share of the inheritance would be zero. He had a prodigal sister.

Chapter Nine
Career Guidance

First part: Problems from this chapter

109. Lucy had her third birthday. She asked her mom to cut her birthday cake in the shape of a locomotive. Her mom wasn't really sure what a locomotive looked like. She did the best she could.

    The wheels were made out of marshmallows.*

    She got eight more dolls and a gallon of perfume, and said, "Yucky!" Lucy knew that she would use the dolls as passengers. She poured the perfume in the engine and pretended that it was diesel fuel for her diesel locomotive.

    Pouring a gallon of perfume into a toy engine meant that almost all of it ran out and onto the living room carpet. It stunk. She declared it an environmental disaster area and moved all her train stuff into her bedroom.

    Lucy wasn't interested in kid stuff.

        She wasn't interested in eating birthday cakes.

            She wasn't interested in hopscotch.

Her only passion was trains.

    She ate so she could have energy to work on the Lucy Railroad.

    She slept so that she could be more awake to think about trains.

    She read train books.

    She looked on the Internet to see if there were colleges that offered an M.T.A. degree. (Masters in Train Administration) The closest she could find were colleges that offered an M.B.A. (Masters in Business Administration)

    Before you start a business, you should have spent at at least 5,000 hours learning about that business. That's the very minimum. Many businesses fail because their founders haven't done the necessary learning.

    Lucy was spending ten hours each day immersed in all aspects of railroading. Roughly, how long will it take her to amass 5,000 hours of learning?

---

* Almost everyone misspells this word.

301. Lucy was playing with her trains in her bedroom. Her mother brought her a piece of birthday cake. Lucy had completely forgotten that the family was celebrating her birthday.

She popped one of the marshmallow wheels into her mouth, said "Toot!" and continued playing. *Toot!* was Lucy's word for *thank you*. It was also her word for *good morning* and for *please pass the salt*. When you only can speak four words, they had to have lots of meanings.

This was especially true since Lucy's first three words—yucky, locomotive, caboose—each had only one meaning.

If the domain is {toot} and the codomain is {thank you, good morning, please pass the salt}, is this a function? *Assign to toot to thank you; assign toot to good morning; assign toot to please pass the salt.*

338. On Lucy's third birthday her spoken vocabulary expanded to five words: {yucky, locomotive, caboose, toot, isn't}.

Locomotive, caboose, and toot are nouns. Nouns are persons, places, or things. Toot is a thing. It is the sound that a locomotive makes.

Yucky is an adjective. Adjectives are words that describe nouns.

Isn't is a verb. Verbs describe what nouns do.

Lucy could now make sentences.

Locomotive isn't yucky.
Caboose isn't yucky.
Locomotive isn't caboose.
Caboose isn't locomotive.

Lucy's mom ran to Lucy's baby book and wrote under Baby's First Sentence: *Locomotive isn't yucky.*

You, my reader, know a lot more words than Lucy. Is it possible to make a sentence with no verb in it?

454. $4\frac{1}{5} \times 7\frac{1}{8}$

542. One gallon of perfume that Lucy received for her birthday cost $2.07. (It was cheap perfume.) How much would a liter of that stuff cost?

One gallon ≈ 3.785 liters.

≈ means "approximately equal to"

Second part: the 𝔐ixed 𝔅ag: a variety of problems from this chapter and previous chapters

656. Lucy's dad bought three gas masks so that he, his wife, and his son could stay in the stinky living room. The total cost, including the $3.98 shipping charge, was $25.82.

How much did each gas mask cost?

771. When they put on their masks, they could still smell the perfume. Dad had accidentally ordered Halloween masks instead of gas masks.

4 ft.

6 ft. 3 in.

"Normal kid" Luke sprang into action to repair the damage that "train kid" Lucy had done.

He cut out the rectangle of carpet that had the perfume spill on it and threw it out the window. Suddenly the living room smelled better.

How many square inches had Luke cut out?

833. Of course, there was now a big hole in the carpet. Dad picked up the phone and called **Cammy's Custom-Cut Carpets**.*

Cammy asked, "How much carpet do you need?"

Dad responded, "3,600 square inches."

Cammy explained that in the United States carpet is usually sold by the square yard.

Help Dad out. How many yards are needed. (Hint: Since there are 36 inches in a yard, there are 36^2 square inches in a square yard.)

---

* If that looks a bit familiar, that's because Cammy and Sammy (see four pages ago) are sister and brother.

Chapter Ten

Money

First part: Problems from this chapter

162. Suppose the government decided that we must use tall trees for money. Only trees over a hundred feet tall could be used. Why would this not be a good choice?

      Recall that a good money system has these five properties: durable, portable, limited supply, fungible, and divisible.

275. The government could get 100,000 toothpicks.
        It could dye some of the red. Those would be 1-picks.
        Some of them blue. Those could be 10-picks.
        Some, green. Those could be 100-picks.
        Some orange. 1,000-picks.
        Some purple. 10,000-picks.
        Why would this be a bad money system?

328. When you don't have a money system, you have to trade (barter). That can really slow things down. Lucy has 36 dolls that she would be happy to get rid of.
        She could trade 3 dolls for 2 hammers.
        She could trade 4 hammers for 5 screwdrivers.
        She could trade 6 screwdrivers for 5 train cars.
Using conversion factors, convert those 36 dolls into train cars.

489. Lucy had a dream about one of her trains carrying 176 people. (People = passengers + train workers)
She knew that each coach car could carry 24 passengers. The train would need 8 train workers. How many coach cars would there be? (Start this problem by Let x = the number of coach cars.)

Second part: the 𝔐ixed 𝔅ag: a variety of problems from this chapter and previous chapters

Check Writing—Lesson Two

Besides not leaving extra spaces (as Fred had done), you shouldn't use a decimal point in writing the dollars.

Lucy was practicing writing checks. She wrote . . .

LUCY RAILROAD   581

Date *May 6*

Pay to the order of   *Train Station Statuaries* $ *64.00*

Sixty-four   *dollars*

Kittens Bank

"IT'S IN THE KITTY"   Lucy

▲▲▲▲▲▲▲▲▲▲▲▲▲▲▲▲▲▲▲▲▲▲▲▲▲▲▲▲▲▲▲▲▲▲▲▲

It is very easy to change that decimal point!

LUCY RAILROAD   581

Date *May 6*

Pay to the order of   *Train Station Statuaries* $ *64,000.*

Sixty-four thousand   *dollars*

Kittens Bank

"IT'S IN THE KITTY"   Lucy

▲▲▲▲▲▲▲▲▲▲▲▲▲▲▲▲▲▲▲▲▲▲▲▲▲▲▲▲▲▲▲▲▲▲▲▲

A check for $64 should be written as . . .

LUCY RAILROAD   581

Date _May 6, 2017_____

Pay to the order of ___Train Station Statuaries_____ $ 64⁰⁰_____

_____Sixty-four and 00/100_____ dollars

Kittens Bank

"IT'S IN THE KITTY"   _____Lucy_____

▲▲▲▲▲▲▲▲▲▲▲▲▲▲▲▲▲▲▲▲▲▲▲▲▲▲▲▲▲▲▲▲▲▲▲

64^{\underline{00}}$ not $64.00   The cents are $^{\underline{\text{raised and underlined}}}$.

547. Solve $6y + 59 = 8y + 10 + 5y$

627. Lucy was looking on the Internet for things to buy for Lucy Railroad.

Getting some train station roof coating was an obvious thing that she would need.
One company offered: 12 cans + $50 shipping.
A second company offered: 20 cans + free shipping

Twelve cans + $50 shipping from the first company would cost the same as 20 cans from the second company. Find out what that price per can is.

774. (continuing the previous problem) A third company had a sale. They offered 19% off their regular price of $7.77 per can. Was that a better deal than the $6.25 per can that the first two companies offered?

Chapter Eleven
Borrowing Money

First part: Problems from this chapter

119. Lucy knew that different train cars are used for different purposes. A coach car is good for transporting people, but it isn't very good for hauling coal.

> Different numbers are used for different purposes.

✳ If you want to count people, you can use whole numbers like 0, 39, or 6,848,715. Numbers like ⅝ are not very useful.

✳ If you are buying yards of fabric in a store, numbers like 1⅝ or 2¼ are useful, but buying π yards of cloth or –4 yards of cloth wouldn't work very well.

✳ If you are looking at a thermometer in the middle of winter in North Dakota, numbers like –25 would be useful. (That's 25 degrees *below* zero.) Numbers like 93 or 650 wouldn't apply in this situation.

✳ The integers, {. . . –3, –2, –1, 0, 1, 2, 3, . . .}, are handy when you are buying stocks. You can own 100 shares of a company. You can own 5,000 shares. You cannot buy 38½ shares.

> Many people don't know this, but you can have –400 shares of a company in your stock market account. It's called **shorting a stock**. It is done by selling 400 shares of a stock *that you don't own*. Later you buy 400 shares and you then have zero shares in your account.

◇ ◇ ◇

> **Please stop for a moment. I, your reader, have never heard of this.**
> I told you that most people don't know about shorting a stock.
> **I don't want to be "most people." Tell me why in blazes would someone sell a stock they don't own and then buy it back later?**
> Most people buy shares of XYZ company at, say, 40, and sell it when it hits 50. They make money. It's called buying low and selling high.

If I think the stock of XYZ is overpriced at 40, I'll sell it at 40 and later buy it back at 30. It's called shorting a stock—selling high and buying low.

◊ ◊ ◊

Now, where were we before you interrupted?

The numbers you use depends on what numbers you need. Your question: Are there numbers that are not on the number line?

Translation: That means are there numbers that are not less than zero, equal to zero, or greater than zero? *Please make a guess.*

248. *While we are in the guessing mode . . .* Is there a number that is larger than every number on the number line? We are not talking about ∞, which just means unboundedly large. Is there an actual number?

Second part: the 𝔐ixed 𝔅ag: a variety of problems from this chapter and previous chapters

332. Three-year-old Lucy knew that she would need money in order to create the Lucy Railroad. No one was going to hand it to her. She wasn't going to steal it.*

The 𝔗raditional 𝔄pproach to making a lot of money is to get a job for several years and save some money. Then use that money to start a business.

Sometimes the business will fail, and you repeat the cycle of working for wages and then start a business a second time.

A second time! I, your reader, don't like the sound of that.

Businesses fail—especially if you haven't spent the hours learning about the business. (Chapter 9)

But what about the shame?

Four US presidents went bankrupt: Lincoln, Jefferson, Grant, and McKinley. (Lincoln went bankrupt twice.)

Rembrandt (the painter) went bankrupt.

Walt Disney went bankrupt.

Henry Ford (the guy who started the Ford Motor Company) went bankrupt.

And many, many more.

**Failure only happens
to those who attempt
to succeed.**

The rest can work for wages, if someone will hire them.

---

* When Lucy was 1, she did steal stuff. (Chapter 2) But she is older now and never again will take what other people own. She had read about C.C. Coalback and thinks that he is yucky.

Chapter Eleven   Borrowing Money

Lucy was going to try the Less Traditional Approach.

At age 3 or 4 or 5 or 6 or 7 or . . . or 14 or 15 or 16, you really can't make much money working at a job.

Instead, she would spend those years getting ready to start the Lucy Railroad. She would read books, magazines, Internet articles. She would talk with people who have been successful in business. She would talk with anyone connected with railroads—passengers, locomotive engineers, conductors, other railroad owners. She would talk to people who are good at advertising—an important part of most successful businesses.

And she would talk with people who lend money. (See the title of this chapter.) Lucy would find out all the aspects of borrowing money:

✓ How much money could she get.

✓ What interest rate would be charged.

✓ How long could the money be borrowed for. (You wouldn't want a loan that was due in only one year. It takes longer than that to get most businesses started.)

✓ What collateral would the lender want.

With only five words in her spoken vocabulary, {yucky, locomotive, caboose, toot, isn't}, Lucy decided that interviewing all those people would have to wait until later. In the meantime she would read books.

If you are 3 years old, what percent increase is needed to become 16?

463. Lucy understood everything that was said to her. She was an excellent reader. She could do arithmetic ("19% off the regular price") and elementary algebra ("12 cans of roof coating").

She had even started reading in an advanced algebra book about some short kid who was on a two-day trip home from his time in the United States army.* When she got to the imaginary number i, she decided to wait until she was 4 before she continued reading that book.

Instead she read Prof. Eldwood's *Modern Approaches to Speech Pathologies*, 1847. He listed a zillion reasons why a kid might have a very small speaking vocabulary.

_____

* Hint: *Life of Fred: Advanced Algebra Expanded Edition.*

Chapter Eleven   Borrowing Money

Eldwood wrote that there might be physical problems or mental problems or emotional problems that might be getting in the way.

Lucy thought to herself Hogwash! I haven't been talking much because there is no need to rush things. I can order things on the Internet by typing. I can read books. I am not dumb, but I have chosen to be almost dumb.[*]

It was one week after her third birthday. Lucy decided to speak. Instead of saying just *yucky*, she had a zillion ways to express that concept: detestable, contemptible, disgusting, repugnant, repulsive, revolting, gross, foul, filthy, odious, obnoxious, abominable, abhorrent, hateful, execrable, heinous, and deplorable.

She had never needed a phone before, but now it was a necessity. She ordered the Super Business Phone from the Ding-a-Ling phone company.

Special telephone poles were installed to handle the 16 phone numbers that Lucy needed.

The normal installation charge was $1,036, but Lucy needed everything in a hurry. She chose to pay the extra 24% fee for fast service.

How much was the total bill?

516. That afternoon Lucy's mom noticed a lot of construction noise. Workers would installing telephone poles on their front lawn. She ran out and asked what they were doing. The foreman showed her the work order.

Ding-a-Ling Phones
Work order number 9873085
Install 16-line service
Customer name:
    the Lucy Railroad company

The monthly service fee is $64.48 plus 52¢ per call.

In the first month the bill was $130. How many calls had Lucy made?

---

[*] *Dumb* has two meanings. Lucy is not dumb (= stupid), but Lucy has been dumb (= not speaking).

Chapter Twelve
Ownership

First part: Problems from this chapter

163. Lucy dreamed of creating the Lucy Railroad. With years of work she would make something that was new to the world. It was her baby.

She didn't want to live in a country that would take her creation away from her. Lucy thought that socialism was more than yucky; it was odious.*

Lots of people like socialism. Make a guess why they do.

230. In a pure democracy 51% of the people can make any laws they like. They can make laws that are hard on the Irish or the Chinese, or the productive members of the society. Any minority is vulnerable to the will of the majority.

The Declaration of Independence was the document that said that we were leaving the British government. ". . . [W]henever any form of government becomes destructive of these ends [the rights of life and liberty], it is the right of the people to alter or abolish it, and to institute new government. . . . ."

The Constitution of the United States was the document that created our present government.

Make a guess . . . Which of these two documents has the word *democracy* in it?

    A) the Declaration of Independence
    B) the Constitution
    C) both of them
    D) neither of them

392. If x is an even number, what is the next consecutive even number?

_____

* With Lucy's new spoken vocabulary, she could more accurately describe things. *Yucky* might mean only distasteful. *Odious* (OH-dee-us) means highly offensive or disgusting.

Second part: the 𝔐ixed 𝔅ag: a variety of problems from this chapter and previous chapters

486. If x is an odd number, what is the next consecutive odd number?

517. If two consecutive even numbers add to 1,294, what are those numbers?

      Begin by writing   Let x = the first consecutive even number.

653. The Pareto* principle deals with the productive members of any group. It is sometimes called the 80/20 rule.

      20% of salesmen make roughly 80% of the sales.

      20% of railroads do roughly 80% of the shipping.

      20% of employees make roughly 80% of the suggestions.

      20% of golfers win roughly 80% of the awards.

      20% of movies will contain roughly 80% of the violence.

      This principle was discovered by Vilfredo** Pareto in 1896 when he noted that 80% of the land in Italy was owned by 20% of the people, that 80% of the peas in his garden were produced by 20% of the pea pods.

      When the Lucy Railroad gets large and has 175 employees in its home office, roughly how many of them will do 80% of the work?

776. What gets really exciting is if you look at the 20% who are doing 80% of the work.

      Let's apply the Pareto principle to those 20%.

      This will say that 20% of those 20% will be doing 80% of the 80% of the work.

      Simplify the math in the previous sentence.

-----

* *Pareto* is pronounced either pur-REE-toe or pur-RAY-toe.

** It is just by chance that *Vilfredo* has *Fred* in it.

Chapter Thirteen

Interest Rates

First part: Problems from this chapter

165. When Lucy was older and she started the Lucy Railroad, there was no way of predicting whether it would be successful. If it were successful, she had no way of predicting how fast it would grow.

She imagined . . .

What if I were 18 years old when I finally could get my first engine and train car. Suppose my total initial investment were $6,000. And suppose the Lucy Railroad would grow at the rate of 2% per month.

How much would she have at the end of one month?

228. . . . at the end of 4 months?

313. Interest can be happy or unhappy. It can be happy in two ways and unhappy in one way.

#1 Happy: In the previous two problems the Lucy Railroad was growing at 2% per month.

#2 Happy: When you borrow money to buy productive assets. If she used her mother's Disaster credit card and didn't pay the monthly bill, then Disaster would charge 1½% per month. But if she buys things she needs for her Lucy Railroad, then she would be making 2% on that borrowed money. This would be a happy situation.

Your question: 1½% per month times 12?

453. #1 Unhappy: If you use your Disaster credit card to buy things that aren't productive *and* don't pay the balance due on the monthly bill, then that charge of 1½% can hurt you financially.

It is soooooooooooo easy to buy things using a credit card. It doesn't seem to hurt the way that paying cash does. You just sign your name and the store hands you goodies.

It's not unusual to see a credit card balance of $12,000. What is 1½% interest charge on that balance?

Second part: the 𝔐ixed 𝔅ag: a variety of problems from this chapter and previous chapters

544. Right now, at the age of 3 years and one week, Lucy knew that one of the most productive assets she could get was education. The 18 statues and the iron railing that she had purchased and put in storage might someday be useful, but right now it was more important to stuff her brain than stuff her storage units.

Books were cheap, especially compared with college.

Going to train stations and switching yards was free. (Her mom drove her.)

Talking with other railroad owners, with accountants, with lawyers, with advertising agents—these all gave her valuable information.

She realized that she didn't need 16 phone lines. Two lines would be plenty for her right now. What percent decrease in the number of phone lines would that be?

647. In many accounting books Lucy was told that there were two important things to do:

↑ Increase your income
↓ Decrease your expenses.

That made a lot of sense to her. Each of **Sammy's Stuff-It-In Storage Units** was costing $125 per month.

Prof. Eldwood's *Modern Art of Deal Making,* 1847, gave her an idea. She said to her brother Luke, "How would you like to make $120 a month?

Luke was surprised. He didn't know that his kid sister could talk. He asked, "How?"*

Lucy explained, "You know those 18 statues I have in one of my storage units? I

---

* Not all sentences need to have a verb.

give you $120 if you let me store them in your bedroom."

Luke smiled and said, "It's a deal."

Lucy's monthly storage bill, which was $500 (= 4 × $125), now became $495 (= 3 × 125 + 120). Which percent decrease was that?

       3 × 125   the three storage units Lucy still kept

          120   the payment to Luke

Chapter Fourteen
The Tragedy of the Commons

First part: Problems from this chapter

Lucy's mom looked out her front window and noticed that the telephone workers had cut the lines to their house. She asked Lucy what was going on.

Lucy said that she was learning to save money. Instead of the 16-line phone service, she had switched to a 2-line cell phone. Lucy said, "I'm trying to cut expenses."

Her mother thought Thank Goodness. That was a prayer of thankfulness.

She then asked Lucy, "What about those three telephone poles in our front lawn?"

Lucy shrugged her little shoulders. "That's okay, Mom. I don't mind them." What Lucy was really saying is that she didn't want to bear the expense of removing those poles.

That front lawn was "owned" by the four members of the family. When Lucy had the telephone poles installed, she was replaying The Tragedy of the Commons. What is owned in common is subject to overuse by each of the owners.

122. Let's look at it in reverse. Suppose it cost $1,000 to remove those poles. How much should each family member pay?

252. The wealth of a community is owned by the individuals (or families) who have created that wealth. If Lucy's father works in a tire store, he trades his labor for a paycheck.

Governments often treat all the money in the community as if it were owned by the community collectively. It takes (taxes) money from the individuals and uses it for the "common good."

It would be fair if: ① each person was taxed the same as every other person, and ② everyone received the same benefit from the government spending.

Which of these two (① or ②) is not true?

Second part: the 𝔐ixed 𝔅ag: a variety of problems from this chapter and previous chapters

390. Lucy learns 7 new railroad ideas in every 12 minutes from the reading that she does. In 88 minutes how many ideas might she expect? (Round your answer to the nearest whole number.)

452. Lucy has a collection of 2,307 cards. On each card she has written down one idea that she has gotten from her reading.

Let those cards be the domain.

She has 18 boxes, each with a label on it. One box, for example, says Accounting Ideas. Another says Advertising Ideas. Another says Miscellaneous Ideas that don't fit in any of the other 17 boxes.

The codomain is those 18 boxes.

Lucy puts each card in a box. This is a function.

Your question: Is this function one-to-one?

545. If x is an odd number, what are the next two consecutive odd numbers?

650. The sum of three consecutive odd numbers is equal to 1,647. What are these numbers?

773. The city street in front of Lucy's house is an example of the commons. If you throw your trash on the street, then the city cleans it up and increases everyone's taxes by a little bit to pay for the extra cleaning.

If everyone did that, the city streets would become a dump.

What would stop Lucy from cutting down those three telephone poles and rolling them into the street?

830. $5\frac{1}{5} - 2\frac{1}{3}$

Chapter Fifteen
Fred Finds His Business

First part: Problems from this chapter

145. Lucy turned 4. She was really getting old now. Her mom bought her a lovely doll house so that she could learn to "play house."

Lucy squealed, "Thanks Mom. Finally you have given me something I like."

She spray painted it brown to make it look more like a train station. When she took it to her room, she put some of the dolls around the train station and pretended that they were passengers.

When she looked inside the doll house, she found that it was filled with eye shadow and false eyelashes. She was disgusted. She threw them away and filled the train station with little vending machines. My passengers have got to eat she thought to herself.

Originally, she was thinking of putting a Lucky Lucy's Lounge in her train station. There would be a band playing quiet music, and a restaurant featuring BBQ ribs, hamburgers, spaghetti, burritos, chow mein, and six flavors of Sluice. There would be a three-screen movie theater and a skating rink and a shoe store.

Forget that she thought and stayed with the vending machines.

Your question: What would be the drawbacks for Lucy if she really did put all that extra stuff in her train station?

274.   Fact #1: Even something as simple as opening a flower shop requires a zillion different skills: knowledge of flower arranging, of locating businesses, of advertising, of hiring people, of ordering flowers (you are not going to grow them yourself!), of shipping, of handling credit card sales, of handling lawsuits, of dress codes for employees, and so on.

Fact #2: If you find people who are educated in each of these particular areas and if you form a big partnership with everyone having a share in the ownership and the profits, then you will face The Tragedy of the Commons. With few exceptions, partnerships don't work.

Given these two facts, what should you do?

Second part: the 𝔐ixed 𝔅ag: a variety of problems from this chapter and previous chapters

Lucy felt very blessed. At the age of 4 she knew what direction her life was headed: the creation of the Lucy Railroad.

During the first two years of college, students are often required to take courses in many different areas. Those are called the breadth requirements.

You will probably take
        courses in math and science
        courses in the humanities
        courses in the social sciences
        courses in reading and writing English
        and maybe courses in a foreign language.

Often at the beginning of their third year of college, students will select their major. With experiences in a variety of subjects, students will be better able to select a major that best fits them. Some big universities offer over a hundred majors.

331. When Lucy is alone in her bedroom, she reads books and studies offerings on the Internet. Many universities now offer the content of many of their courses online at no charge. Without paying, you can learn all the material. You just don't get college credit.

Which one of these would be hardest for Lucy to learn using only books and the Internet?
    A) U.S. history
    B) great literature
    C) chemistry
    D) writing good English
    E) philosophy

461. Lucy saw an ad on the Internet for toy vending machines. She thought This will be perfect to use in my toy train station. I can learn where to place them in the station so that they will get maximum use. Seven machines plus $5.09 for shipping will cost $55. How much is each machine?

Chapter Sixteen

Competition

First part: Problems from this chapter

136. Lucy was designing a new train route. The existing route from Rose to Ziptown was 5 miles long. From Ziptown to Dawn was 12 miles.

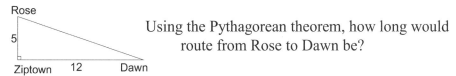

Using the Pythagorean theorem, how long would route from Rose to Dawn be?

201. Lucy wants her railroad to be a super success. In a free market where there is no government interference:

The key to lots of sales

is to please your customers.

The business owner is slave to the people he (or she) serves.

Lucy started a list of how to please her railroad customers:

1. My train stations will be neat and clean.

2.   <<Help Lucy make her list.>>

311. Lucy's train from Reno to Sacramento will travel at 30 mph. When it goes back from Sacramento to Reno it will go 20 mph since it will be traveling uphill and will take 2 hours longer. How long will it take to go from Reno to Sacramento?

   Using "Let t = the time from Reno to Sacramento" and "Then" statements, set up the equation. (In the next chapter, we will learn how to solve that equation.)

487. Use the "six pretty boxes" approach on the previous problem to find the equation.

Second part: the 𝔐ixed 𝔅ag: a variety of problems from this chapter and previous chapters

549. If the legs of a right triangle are 6 and 8, what is the length of the longest side?

Terminology: The legs of a right triangle are the two shorter sides. The longest side of a right triangle is the hypotenuse. (pronounced high-POT-en-noose)

601. Lucy's parents were thinking of adding another bedroom to their 1600 square foot house. It would be 10' × 12'. (ten feet by twelve feet)

What percent increase in area would that be?

703. Lucy was overjoyed. She ran around the house shouting, "Yes! Yes! Yes! Yes! Yes! Yes! Yes! Yes!" Even though she had a large speaking vocabulary, in this case a single word was enough.

Lucy thought I can use this bedroom to set up a big toy train layout. I can buy toy tracks, switches, toy train stations, and a transformer to power the tracks.

how Lucy pictured the new bedroom

Luke thought Great! I can get those 18 statues out of my bedroom.

Lucy's mom thought In five months Patrick or Patricia will be here. Right now I'll call him/her Pat.

Chapter Sixteen   Competition

If the domain is the thoughts of these three people {train set, statue removal, baby} and the codomain is {☺, ☹}, create a function and tell whether the function you created is one-to-one.

900. Lucy saw a bunch of old women (in their 20s and 30s) come and give a party for Lucy's mom. They gave Lucy's mom a whole bunch of gifts: a bunch of cloths, a bottle of talcum powder, a rattle.

Lucy thought this was weird. You build a new bedroom on your house and people give you funny presents? I wonder what they'll give me when I build 18 train stations? Luke explained to Lucy that this was called a baby shower. Lucy began to hope for train station showers some day.

The cloths came in a package marked Ralph's Diapers. Lucy took 6 diapers and experimented. With 6 diapers she could clean the oil off 8 of her toy engines.

How many diapers would it take to clean the oil off of 20 engines?

Chapter Seventeen
Killing Competition

First part: Problems from this chapter

168. Lucy had her fifth birthday party. Patricia had moved into the new bedroom.

Both Lucy and Luke were jealous of all the attention that their mom paid to Patricia. Mom was always in the new bedroom diapering Patricia.

Lucy asked her mom, "Why didn't you get a baby that was already potty trained?"

Mom said, "I'll explain it to you when you are older."

Lucy didn't like that answer.

Lucy and Luke realized they had competition for Mom's attention and they didn't like it.

The first method to kill competition is collusion. They asked each other how they could get more time from their mother.

Lucy suggested that they start wetting their pants. Luke thought that idea was stupid. "Besides," Luke explained, "at our age she would make us change our own pants and put the dirty ones in the wash."

Luke had an idea. "We could fight. Then Mom would have to come and break it up."

Lucy didn't like that. "But we never fight. You know that I love you almost as much as I love railroads."

Luke took that as a compliment.*

Luke said, "Okay. Let's start fighting."

---

* Not to be confused with *complement.*
Compliment = to say nice things.
Complement = to complete, to put things together to make a whole.

In geometry two angles are complementary if they add up to a right angle (= 90º).

In music complementary intervals add up to an octave.

Lucy just stood there. She hadn't watched as much television as Luke and wasn't sure how to fight. She asked, "What are we supposed to do?"

Luke said, "On television and in the movies people say nasty things to each other. That's fighting."

Lucy shrugged her shoulders. "That's it? That's what fighting is all about? I don't like that."

Luke said, "Give it a try."

Lucy looked a Luke for a long while. She was thinking. She said, "You are a ping-pong ball."

Luke gave up.

If they wanted to get more of their mother's attention, they would have to be C & A*, rather than a pain in the neck.

When their mom headed into Patricia's room, they followed her. Lucy begged her to show her how to change a diaper. Luke said, "Show me too." This was pure collusion. Mom would have to be teaching her two older kids rather than paying attention to baby Patricia.

They worked for a total of 5 hours.
If Lucy worked for t hours, how many hours did Luke work?

250.   Lucy could change 5 diapers per hour.
Luke could change 8 diapers per hour.
In 5 hours they changed a total of 34 diapers.
How many hours did Lucy work?
Begin by "Let t = the number of hours that Lucy worked."

312. Do the previous problem using six pretty boxes.

460. In the previous chapter (problem #311), we learned: Lucy's train from Reno to Sacramento will travel at 30 mph. When it goes back from Sacramento to Reno it will go 20 mph since it will be traveling uphill and will take 2 hours longer. How long will it take to go from Reno to Sacramento? We arrived at the equation $30t = 20(t + 2)$. Solve this equation.

_____

* Cute and Adorable

Second part: the 𝔐ixed 𝔅ag: a variety of problems from this chapter and previous chapters

571. After Lucy and Luke had changed the sixth diaper, they realized that their plot to get their mom to play more attention to them was failing. Aftcr changing six diapcrs, thcy kncw how to do it without any supervision. Mom was no longer in the room to watch them.

Is *sixth* a cardinal number?

Is *six* a cardinal number?

651. Lucy's mom made her final entry into Lucy's baby book. Under Baby Is Growing Up she wrote: *My little choo-choo lamb is no longer a baby. She and her brother have both been helping change Patricia's diapers. They are both so sweet.*

She closed the book and tied a ribbon around it. She planned to give it to Lucy when she turns 18.

During the first 24 months of Lucy's life, her mom had made r entries into Lucy's baby book each month. In the second 24 months of Lucy's life, she made entries at the rate of r – 6 per month.

There were 1,296 entries at the end of those 48 months. Using six pretty boxes, find the value of r.

732. At the age of 5, Lucy thought that being called "my little choo-choo lamb" was too babyish. She thought of herself as the Queen of Railroad. She cut a picture out of a magazine and showed it to her mom.

"I want you to think of me this way," Lucy said. "I have a great future in the world of railroads. I will be the Lucy, the Queen of Railroad. You will be the mother of the Queen."

For Lucy, her thoughts were always around the Lucy Railroad. At 3, it was in order to make a lot of money. $$$$$$$$$$$$

Now, at 5, it was to be famous . . . like a queen.

When she grew up and became 7, the juvenile thoughts of money and fame would fade.

A hard question, which many adults would have trouble answering: Beyond money, beyond fame, what is there to achieve?

Chapter Eighteen
A Second Way to Kill Competition

First part: Problems from this chapter

114. Lucy had her seventh birthday party. Patricia was now 2, so Lucy and Luke didn't have to change her diapers anymore.

     In the roughly 700 days since her fifth birthday, Lucy had been reading books about railroads. She had talked/phoned/written/emailed to everyone she could think of who might give her help in her Great Work: to establish the Lucy Railroad.

     Lucy was focused on her Great Work.

     She kept a diary of all the things she learned about railroading.

     Each week she wrote r ideas down about the costs of buying railroad things—cars, stations, land. . . .

Each week she wrote down $r + 5$ ideas about what her customers would like to have in dealing with Lucy Railroad. (If you want to succeed in business, your customers' desires are at least as important as finding ways to fill those desires.)

     After 20 weeks Lucy had a total of 580 ideas in her diary. How many ideas per week did she write down regarding buying railroad things?

     Solve this using six pretty boxes.

218. How many ideas per week did she write down regarding her customers' needs?

336. Lucy slept well. She ate well (no cupcakes). She flossed and brushed her teeth. She exercised. Being in **good physical health** is one key part of accomplishing almost anything important in life.

     Which of these would slow down Lucy's progress?

        a. Being in jail.

        b. Watching television all day long.

        c. Spending her savings on "fun stuff" such as fancy shoes, comic books, cars, jewelry, vacations, and candy.

        d. Becoming a bull fighter.

        e. Waiting until she was 65 before starting to work on creating the Lucy Railroad.

Second part: the 𝕸ixed 𝕭ag: a variety of problems from this chapter and previous chapters

473. Lucy had studied the maps of existing railroads in the United States. There was so much that she wanted to learn by the age of 16. She was now 7. How many hours will it be before she turns 16? (Use 365 days = 1 year.)

550. Few seven-year-olds are as focused as Lucy. Other kids head off to kindergarten where they play hopscotch and beat on tambourines.

few people can spell tambourine

    In the fifth grade they learn about birds. In high school they go to football games and proms.

    For Lucy this whole thing would be a giant waste of her time.*

    She heard that some kids drop out of high school at 16. She didn't want to wait 78,840 hours. She announced to her mother, "I am going to be the world's first kindergarten dropout!"

    Her mother laughed. "You dummy. Kindergarten usually starts at about age 5. You are 7. We have been home schooling you for the last two years. You are not in the government school system."

    Lucy had learned so much more than 99% of the government-schooled kids. For what Lucy wanted to do with her life—create the Lucy Railroad—there were two important academic subjects. Pick those two out of this list:

    i) French
    ii) music, including tambourine playing
    iii) social studies
    iv) English
    v) United States history
    vi) Roman numerals (which have been used to make this list)
    vii) math

---

* It isn't a waste of time for many kids. They need to beat on tambourines and go to proms in order to finally figure out what they will need to have for the half dozen jobs that will occupy their years until 65. During those years on the job they will think about their high school football games and prom and will look forward to retirement when they can play BINGO on Thursday nights.

Chapter Nineteen
A Third Way to Kill Competition

First part: Problems from this chapter

164. Two chapters ago we saw how companies could collude to set higher prices. They didn't have to compete with each other if *they all ganged up against the customers.* They would all agree on artificially high prices.

One chapter ago we saw how companies could kill their competition by lying about other companies. Politicians do this a lot.

In this chapter we described a third way to eliminate competition. What was it?

"But Mommy!" Lucy exclaimed. "If I don't go to the government school, won't I be behind all those other kids?"

Her mom explained, "Being a kindergarten dropout—as you call it—has put you way ahead of the pack. Besides learning all that stuff about railroads, you know more English and math than kids who are twice your age."

She told Lucy to put away her calculator and do these:

FIRST-GRade TeST

302.
$$839 + 764 \qquad 3247 - 1889$$

SeCONd-GRade TeST

309.
$$78 \times 64 \qquad 869 \times 70$$

THIRd-GRade TeST

337. $84)\overline{23184}$

FOURTH-GRade TeST

402. $\dfrac{3}{5} + \dfrac{2}{3} \qquad \dfrac{7}{8} - \dfrac{1}{3}$

FIFTH-GRADE TEST

466. $\dfrac{3}{4} \div 2\dfrac{2}{5}$ $8\dfrac{1}{8} - 4\dfrac{1}{4}$

SIXTH-GRADE TEST

471. $0.07\overline{)28.63}$
        Twelve is what percent of 18?
        What is 37½% more than 24?

        Lucy told her mom, "This math stuff is so cinchy."
        Her mom said, "You are only 7 and your doing sixth grade math.
You aren't missing anything by being a kindergarten dropout."
        "But Mom! What about English?"
        She handed Lucy the English quizzes.

FIRST-GRADE TEST
491. Spell cat.

SECOND-GRADE TEST

515. If you are writing a news report about Lucy being a kindergarten
dropout, there are famous question words every reporter uses. Five of
them are: When, Who, How, Why, and Where—4 W's and one H.
        What is the sixth famous question word?

Second part: the 𝔐ixed 𝔅ag: a variety of problems from this chapter and previous chapters

THIRd-GRadE TEST

556. The past tense of *sing* is *sang*. Today I sing. Yesterday I sang.
The past tense of *smile* is *smiled*. Today I smile. Yesterday I smiled.
What is the correct spelling of the past tense of *panic*?

FOURTH-GRadE TEST

646. Find three errors in this letter.

> Dear Luke,
>
>   Thank you for being my brother. I will always
> be hear for you if you need me.
>
>   Yours Truly
>   Patricia

FIFTH-GRadE TEST

654. There are over a half million words in the English language. The average adult uses about 20,000 of them in everyday speaking. I can think of only 24 words that do not contain a vowel (*a, e, i, o, or u*).

Try to name at least two of them before you look at my list in the answer.

SIXTH-GRadE TEST

901. Which of these is correct?
          A) Yesterday I dreamed of pizza.
          B) Yesterday I dreamt of pizza.

Which of these is correct?
          A) Yesterday I kneeled before the King.
          B) Yesterday I knelt before the King.

Chapter Twenty
A Fourth Way to Kill Competition

First part: Problems from this chapter

167. Lucy would spend an hour each day studying math and an hour studying English. Another eight hours she would spend on railroads—reading about, talking about, thinking about, dreaming about trains.

  That was from 7 a.m. to 5 p.m each day, Monday through Saturday. After 5 p.m. she would goof off and do things like eat.

  How many hours of "train time" would she accumulate in 52 weeks?

200. Lucy would go hiking in her neighborhood to see where train tracks might be laid. That gave her good exercise and taught her to look at land in a different way than most seven-year-olds.

  It was cold on the first part of one long hike. She drank one cup of water per hour. The second part of the hike was warm, and she drank 3 cups of water per hour. It was warm for 4 more hours than it was cold. She drank a total of 16 cups of water. (Amazing girl!) How long was it cold? Use six pretty boxes to solve this.

249. Later that week she took a second hike. She hiked 2 hours in the forenoon and 3 hours in the afternoon. (Some people call the forenoon the morning.)

  At noon she had an anchovy pizza with salty popcorn for desert. She drank x cups of water per hour in the forenoon and x + 2 cups of water per hour in the afternoon. (A lot of salt can make your thirsty.)

  She drank a total of 12 cups of water on her hike. In the forenoon how many cups of water per hour did she drink?

Second part: the 𝔐ixed 𝔅ag: a variety of problems from this chapter and previous chapters

570. On Lucy's last hike she took the north route for t hours, drinking 2.2 cups of water per hour.

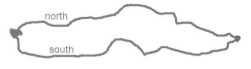

Coming back home on the south route she took two hours longer and drank 1.2 cups of water per hour.

She drank the same amount of water on each route. How long did she walk on the north route? (Translation: Find t.)

648. On her hike she spotted six train stations, which she called A, B, C, D, E, and F. She assigned each of those train stations to one of three categories: ① I'd like to buy it; ② I might buy it; and ③ I would never buy it.

If the domain = {A, B, C, D, E, F} and the codomain = {①, ②, ③}, will this function be 1–1?

704. Before the end of this book Lucy would buy all six train stations. "I never saw a train station that I wouldn't like to own."

Four years after she owned all these stations the net monthly income from these stations was . . .

$300 from station A

–$517 from station B

$889 from station C

$1446 from station D

–$266 from station E

–$49 from station F.

Without using your calculator find the total income from these six stations.

Chapter Twenty-one
Government Regulation

First part: Problems from this chapter

121. Actually, it is better to look at a train station before you buy it. Buying it first and then seeing what you have bought can really hurt you financially. When Lucy said, "I never saw a train station that I wouldn't like to own," she wasn't being very smart.

When she read Prof. Eldwood's *Modern Train Station Buying*, she learned that those two things—looking and buying—were not commutative.

Which of these pairs is commutative?
A) Owning a train station
B) Painting that station
    A) Getting off the train
    B) Going inside the station
        A) Using a vending machine in the station
        B) Using the restroom in the station
            A) Owning a train station
            B) Selling that train station

232. On Lucy's hikes she found the "most adorable" train station. She didn't want to wait "a million years until she was 16" in order to buy it.

She showed a picture of it to her brother Luke. He said that he didn't have any money to help her. He did indicate that if she bought it, he would be happy to relocate those 18 train station statues that were currently in his bedroom out to that station.

She showed the picture to Patricia and she said, "Choo-choo house." Clearly, she was too young to appreciate the financing of the purchase of a train station.

She talked to her parents. Her mom said, "Why would you want to buy a train station? You are only 7 years old. What would you do with it?"

Lucy was flabbergasted. "But Mom! Next year I will be 8! I need it! It will be the first piece of real estate on my way to creating the Lucy Railroad!"*

She headed to her bedroom and closed the door. She didn't want to have anyone see her crying. Creators of railroads and other large businesses never cry she thought. She was wrong. People who dare to dream big dreams will cry a lot more than people who live humdrum lives, but they will often have much more joy in their lives also.

She thought of her Grandma Amy who had created a big business: **Mammy's Mighty Machine Rentals**. Lucy knew that Grandma Amy would understand what it is like to found a business.

Lucy grabbed a cylinder of pennies that she had saved to show Grandma Amy how serious she was. The cylinder was 10 inches tall and had a radius of 2.4 inches. What is its volume? Use 3.1 for π.

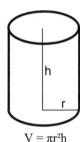

$V = \pi r^2 h$

339. If there were 18 pennies in each cubic inch, how many pennies were in Lucy's cylinder? (Assume the cylinder was filled. Round your answer to the nearest penny.)

472. Lucy phoned.

Voice: "Mammy's Machine Rentals. This is Justin speaking."
Lucy: "Hi. Can I speak to Grandma Amy."
Voice: "Grandma Amy? Oh, you mean Amelia. Here she is."
Amy: "Hello Luce." (Amy was the only one who called Lucy by her pet name Luce.)

---

* One mark of poor writing is the overuse of exclamation points. I mean it!!!!!!!!!!!!

Chapter Twenty-one   Government Regulation

Lucy: "I've gotta see you soon about a real estate deal I'm working on."

Amy: "Sure. Let's meet on the bottom floor of my building in a couple of minutes."

Lucy: "Thank you!"

They saw each other at the end of a 400-foot hallway. They walked toward each other. Lucy was walking 4 feet per second faster than Grandma Amy. In 25 seconds they met and hugged.

How fast was Grandma Amy walking?

Second part: the 𝔐ixed 𝔅ag: a variety of problems from this chapter and previous chapters

580. Lucy was so excited. She motor-mouthed: How she had been out hiking; how she saw this adorable train station; how she was enamored about railroads; how she had to have this station; how her brother didn't have any money to lend her; how Patricia wasn't any help; and how her mother thought this whole idea was nonsense.

Grandma Amy just listened. Lucy showed her the photograph she had taken of the train station and asked, "Well, Grandma. What do you think?"

Amy looked at the picture. "Isn't that the old Bedford station at the north end of town?"

Lucy's eyes lit up. "Yes. That's the one."

"Luce, do you realize that that track is no longer used? It was six years ago that the last train was on those tracks."

Lucy knew *everything* about that. Her years of studying railroads was paying off. "Yes, Grandma. It was six years and four months ago. When I buy the tracks and start my railroad company, that Bedford station will be an important part of my overall plan."

Can grandmothers giggle? Amy did. "Granddaughter, do you know who owns that station now?"

Lucy pulled her smart phone out of her jeans pocket. In a couple of seconds she discovered that MMMR Corp. owned the station.

Then it dawned on Lucy. MMMR Corp. = **Mammy's Mighty Machine Rentals** = Grandma Amy.

"You!" Lucy almost shouted.

Grandma Amy nodded and smiled. "I bought it years ago to store extra machinery for my business. When I built our new building last year, I moved all the machinery to the new building. Would you like that station?"

Lucy couldn't speak. She just nodded as fast as she could and thought Yes! **Yes!** *Yes!* Yes! **Yes!** Yes! *Yes!* Yes! **Yes!** *Yes!* YES! and Yes!

Lucy's blood pressure had been 120/80.* It went up 25% upon hearing the good news. What was her current blood pressure?

* 120/80 means that the pressure was 120 when her heart beat and was 80 between the beats.

630. Lucy had had 0 pieces of real estate. She now had 1 piece of real estate. Was that more or less than 1,000% increase in her real estate holdings?

781. Lucy covered Grandma Amy's face with a zillion thank-you kisses. She ran home, headed into her father's workshop in the garage, and grabbed some pink paint, a brush, and a ladder.

In 20 minutes she was at her train station painting over the old Bedford sign.

In an hour the sign had a new look.

She closed the paint can, washed the brush, and hurried back home. After she put her father's stuff back in the garage, she walked into the house singing, "Hi-diddle-dee-dee, a railroad life for me!" ♪ ♫ (Lucy was vaguely remembering the lines from Disney's *Pinocchio* movie.)

Her mom exclaimed, "You are a pink disaster!" Lucy was covered with pink paint. Clothes into the clothes washer. Kid into the bathtub.

Water pouring in at 3 gallons per minute. Water draining out at 1.2 gallons per minute. (Lucy had accidently kicked the plug out.) How long until the 55-gallon tub is filled?

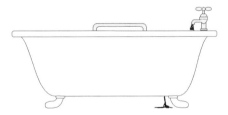

Chapter Twenty-two
Freedom

First part: Problems from this chapter

171. Lucy spent an hour and a half in the bathtub. Her fingers were starting to get all wrinkly from the water. It was the perfect time to think about all the things she would be doing with her first train station. Not many seven-year-olds own a train station.

She hopped out of the tub and dried off. The towel turned pink. In all her excitement about her new train station, Lucy had forgotten to wash her hair. Lucy would have slightly pink hair for the next couple of days.

Back in her bedroom she wrote out her **Action List**. She had read about making an **Action List** in Prof. Eldwood's *The Modern Way to Get Things Done*, 1842.

Making an **Action List** helps you get the important things done. The idea is to divide everything into three categories.

Action List

THINGS TO GET DONE TODAY
    1.
    2.
    3.

THINGS TO GET DONE THIS WEEK
    1.
    2.
    3.

THINGS THAT CAN GET DONE LATER
    1.
    2.
    3.

The whole idea of an **Action List** is that you don't spend your time doing the unimportant stuff.

Under the THINGS THAT CAN GET DONE LATER category, Lucy wrote

1. Get my hair washed.

She filled up her Things to get done today quickly because she had thought of those things while she was in the bathtub. Her brother Luke never thought about creating an *Action List* when he was in the tub. He just liked to play with his toy boats.

First thing on Lucy's Things to get done today list was to raise money. It was time to sell all the yucky stuff that she had received on her birthdays—dolls, fingernail polish, lipstick, perfume, doll house. She placed an ad on the Internet.

Originally, her mom had spent $326 on all that stuff. Lucy figured that she could get 17% of that when she sold it. How much is that?

208. On the Things to get done this week part of the *Action List* Lucy wrote: 1. Open a checking account.

Are these commutative?
  a) getting some money and
  b) opening a checking account

335. Lucy wrote out 43 items on her *Action List*. She also spent 19 seconds scratching some of the pink paint out of her hair. Her head itched. Altogether, she spent 750 seconds on these two things. How long did it take her to write each item?   (Start by "Let x = ...") You may assume that each of the items took the same amount of time to write.

467. $(5\frac{1}{6})^3$ = ?

Second part: the 𝕄ixed 𝔹ag: a variety of problems from this chapter and previous chapters

581. The second thing on Lucy's THINGS TO GET DONE TODAY list was Ask Dad at dinner tonight for garage stuff.

Lucy: "Dad. I've got a woodworking project. Can I borrow your hammer that you have in the garage?"

Dad: "Sure kid. Just return it after you're done."

Lucy: "Dad. Can I also borrow a screwdriver?"

Dad: "Sure kid. Just return it after you're done."

Lucy: "Dad. And some other tools and stuff?"

Dad: "Sure kid. Just return them after you're done."

It would be a while (maybe months!) until she was done with them.

Lucy told her brother Luke that he could put those 18 statues in her train station if he was also willing to do a little hauling for her. Luke said he would be willing to do almost anything to have his bedroom back.

In the morning Luke took 25 trips on his bike and hauled all the tools in the garage and the 18 statues to the **I LOVE YOU AMY STATION**.

With his bike loaded, Luke took t seconds to get from his house to the station. He rode at 20 feet/second.

On the way back home he could go 25 feet/second. The trip home took 72 seconds less than the trip to the station.

Using six pretty boxes, find out how long it took Luke to get from home to the station. (Translation: Find t.)

634. Let T be the set {hammer, screwdriver, saw}. Let L be the set of the 25 trips that Luke took to the station {#1, #2, #3, . . . , #24, #25}.

Assign to each member of T the trip on which Luke took that tool to the station. It turns out that . . .   hammer → #1

screwdriver → #1

saw → #2.

Is this a function from T to L? Is it one-to-one?

835. $\dfrac{3\frac{1}{2}}{4\frac{1}{4}}$

Fractions within fractions! I, your reader, have never had that before. Good luck. ☺

Chapter Twenty-three
Liberty

First part: Problems from this chapter

123. Luke had done his 25 trips. All the tools were now at the train station. For the last seven years most of her train work had been intellectual—lots of reading, talking with people, and thinking about the Lucy Railroad. Lucy was now ready to do some physical work.

Lucy told herself that her train station might need a little bit of work before it was perfect.

She wasn't afraid of work. She had done more work in the first seven years of her life than some people do in their entire life.

A little nailing, a little patching, a little painting—Lucy laid out her plans. It would take 5 days to get the whole job done. What fraction of the work would be done each day?

253. If it took her $5\frac{1}{3}$ days to do the whole job, what fraction would be done each day?

340. The station needed a little electrical work. She worked for 6 hours in the forenoon and installed 7 electrical outlets per hour. In the afternoon she worked for 8 hours. That day she installed a total of 90 outlets. At what rate was she installing outlets in the afternoon? Use six pretty boxes.

470. Lucy had brought her sleeping bag so that she could sleep overnight at the station. She was tired after the 14 hours of electrical work, but for Lucy this was all fun. She was fulfilling her dreams. Her brother Luke on the other hand was also tired. He had been playing tag with his friends all afternoon after school.

Luke didn't notice that Lucy hadn't come home. He was just happy being in his bedroom without the 18 train station statues.

Lucy's dad noticed that the garage was devoid of his tools. He knew that she would bring them back after she was done with them.

Lucy's mom happened to notice that Lucy had been gone all day and hadn't shown up for dinner.* She found out from Luke that her daughter was probably at the train station.

At nine at night she took a flashlight and headed to the station. She knew that the station would be dark since there was no electricity there.

She was wrong.

She could see light pouring out all the cracks in the building. She found Lucy tucked in her sleeping bag reading Prof. Eldwood's *Modern Train Station Repair,* 1850.

She kissed her little choo-choo lamb on the forehead and told her to "sleep tight."

Which of these are commutative?

{ shutting your eyes
{ going to sleep

{ dreaming
{ waking up

{ dreaming about locomotives
{ dreaming about pizzas

---

* Moms are often the first to notice when their children are missing.

Second part: the 𝔐ixed 𝔅ag: a variety of problems from this chapter and previous chapters

582. Lucy slept for 8 hours. If she had slept 20% longer, how long would she have slept? Express your answer in hours and minutes.

655. Find the value of n so that $2^n = 30 + 2$.

730. Dad's roll of electrical wire had been 200 feet long.

Lucy had used $162 \frac{1}{4}$ feet of it to wire the outlets. How much was left on the roll?

850. The wire was very stiff and Lucy's hands were small. It took her 2⅓ minutes to unroll each 10 feet of the wire. How long did it take for her to unroll the 162¼ feet?

Chapter Twenty-four
Creating Wealth

First part: Problems from this chapter

146. Lucy was creating wealth as she worked on her train station. After five days she had transformed

   into

Lucy was a hard worker.

Important point ☞ The wealth that Lucy had created was not taken from anyone else. She didn't make any poorer by improving her train station.
  Translation: It is possible for everyone in a society to become rich.

  In fact, if all your neighbors are rich, it is easier for you to become rich. They have money to trade you for your labor or goods.
  One way to stop everyone from becoming rich is to pass a law that restricts poor, uneducated people from trading their labor for money. Luke would probably be delighted to work for $9/hour sweeping the floors in Lucy's train station. Lucy would be delighted to pay him $9/hour for his work, but if the government has a minimum wage law that says that employers must pay employees at least $10/hour, neither Lucy nor Luke are happy.
  Who does benefit from that minimum wage law?

276. At this point in Lucy's life the worth of Lucy's Railroad was increasing at the rate of 8% each month. Each month it was worth 8% more than the previous month.
  Approximately how long would it take for that worth to double?

Second part: the 𝔐ixed 𝔅ag: a variety of problems from this chapter and previous chapters

393. On one of the days that Lucy was working on her train station she put in t hours doing electrical work, $t - 4$ hours doing plumbing, and $t - 2$ hours doing painting.

Electrical work increased the value of her train station by $16/hour. Plumbing, $18/hour. Painting, $12/hour.

On that day she increased the value of the station by $180. How long did she do electrical work? (Translation: Find t.) Use *nine* pretty boxes.

474. Lucy headed off to visit **Lammy's Lucky Lending**. She took her dad along with her because Lammy might be afraid to lend money to a seven-year-old.

Lucy introduced herself and her father. She told Lammy that her father couldn't speak.* She explained that she owned this lovely train station and wanted to borrow money using the station as collateral.

Lammy

> *Intermission*
>
> When you borrow money, the lender wants to be sure that you will pay it back. They often ask for **collateral**.
> If you don't pay them, then they get to keep the collateral.
> When you get a mortgage on your house, your house is the collateral.

Lammy told Lucy that he would lend 70% of the value of the station. Lammy sent someone out to estimate the value of the station (an appraiser).

The appraiser said the station was worth $360,000. How much would the loan be?

903. Would that loan be more or less than a quarter of a million dollars?

---

* That was true since she had told her father not to speak.

Chapter Twenty-five
Tariffs

First part: Problems from this chapter

170. You should have heard the conversation at KITTENS Bank when the $252,000 check was received. It had been deposited at their ATM.

Teller: "Boss, you gotta see this."

Boss: "That's a big check."

Teller: "This is the account that was opened last week by some seven-year-old kid."

Boss: "I wish she were my kid."

The conversation at Lucy's house was a little different.

Mom: "She did what!"

Dad: "I thought it was going to be a joke—her going to **Lammy's** and applying for a loan. I thought that would teach her a little about the real world."

Mom: "This isn't what I planned for her life. When Lucy was 2, it was okay for her to be playing with train sets, but this is nuts."

Her mom bought some jewelry and a couple of pretty dresses for Lucy. She was going to teach her to be a "proper" girl. She marched over to the train station and walked in.

interior of the train station

Lucy's mom was sure that she wasn't in the right place. This couldn't be her daughter's train station.

The receptionist at the I Love Amy station sat behind a large desk. "How may I help you?" she asked.

Lucy's mom had trouble speaking. "I . . . Is this? What? Where's Lucy?"

The receptionist smiled. "I'm assuming you mean Lucille, owner of this building."

74

"Yes. Lucille? Everyone calls her Lucy, even though her given name is Lucille. Could I see her?"

The receptionist said, "She's in a meeting now with the head of KITTENS Bank. She'll be available in a moment."

Lucy's mom took a seat and waited.

If every second seemed like a minute, then how long did the five-minute wait seem like?

341. Lucy came out. "Hi Mom. Do you like the place?"

Her mom didn't answer. She handed Lucy the jewelry and the dresses. She noticed that Lucy still had some paint in her hair. The two dresses and the $8.08 jewelry all cost $45. Using "Let x = . . . " determine the price of one of the dresses. (The two dresses were of equal value.)

After her mom left, Lucy gave the dresses and the jewelry to her receptionist.

492. Lucy now had a train station. Now it was time to seriously look at buying an engine.

The very same locomotive was offered by two different companies.

your first locomotive

Coalback Loco—$50,000. We are located in Kansas.

Freedonia Engines—$30,000 and we offer free shipping to Kansas.

The choice was easy, except that the United States put a tax (a tarriff) on locomotives purchased from companies in other countries. It was a $40,000 tax.

Who benefits from this tariff?

631. Who is hurt by this tariff?

Second part: the 𝔐ixed 𝔅ag: a variety of problems from this chapter and previous chapters

783. Looking up "sugar tariffs" on the Internet, you can find that Americans pay 87% more for refined sugar than the average world price. Besides making candy bars more expensive, it kills U.S. jobs. A government report admitted, "For each one sugar growing and harvesting job saved through high U.S. prices, nearly three confectionery manufacturing jobs are lost." (2006 Commerce Department study[*])

Why does Congress pass laws restricting U.S. citizens from buying sugar from other countries?

A) Members of Congress hate foreigners.

B) Members of Congress hate dental cavities.

C) Although sugar crops are only 2% of the total U.S. crop production, the sugar producers give 33% of the total crop industries' campaign contributions.

834. If a candy bar would normally cost $2.20, how much would it cost with a 87% increase in price?

905. Lucy puts one-third of her time into reading about railroads. She puts one-fourth of her time into talking with people connected with railroading. She puts one-twelfth of her time working on her train station. Out of 24 hours how much is left for miscellaneous stuff such as eating and sleeping?

_____

[*] This was a real study done by the United States government, which showed the bad effects created by the sugar tariff.

Chapter Twenty-six
Opportunity Costs

First part: Problems from this chapter

106. The person with the highest opportunity cost is the . . .
    A) most likely
    B) least likely
          person to take a job working at **Tammy's Tires**?

303. Lucy was spending one-twelfth of each day working on her train station. With her increasing net worth she realized that her efforts were making about $50/hour.

In the old days (when she was 6), it made sense for her to doing that physical work—emptying waste baskets, dusting her trophies,

Lucy's office

straightening her books. When she was 6, her opportunity cost was near zero. Doing *anything* was a step in the right direction.

Now, doing that janitorial work was a mistake. Lucy had read in Prof. Eldwood's *How to Make a Zillion Dollars in Real Estate*, 1853, that when you are investing in your first rental unit, you will probably be doing all the physical work yourself. You will be doing the cleanup, the painting, the landscaping, etc.

He pointed out that one of the biggest mistakes young real estate investors make is that they continue to do "all the little stuff" after they have acquired several large apartment buildings. He called it "a waste of time" since in 1853 the concept of opportunity cost had not yet been discovered.

If Lucy were to offer a job to do that physical work for $18/hour, which of these people might be willing to accept?

A) Lucy's dad who is currently making $26/hour as an assistant manager at **Tammy's Tires**

B) Lucy's mom who is currently making $13/hour working in a candy store

C) Lucy's brother Luke who makes $5/hour whenever he does extra chores around the house.

Second part: the 𝔐ixed 𝔅ag: a variety of problems from this chapter and previous chapters

350. Lucy hired her mom. She knew that her mom could do a better job of straightening out things at the train station than her brother could. Lucy has one-twelfth of her day freed up. Two hours. (= 24 × 1/12)

   She added that to her sleep time. Seven-year-olds need more than eight hours of sleep each night.

   She looked at each of the hours of a day and numbered them. Let set H = {1, 2, 3, 4, . . . , 23, 24}. She looked at her activities of each day. Let set A = {sleeping, reading, talking with people connected with railroading, planning}.

   She assigned each element of H to an element of A.
1 → sleeping; 2 → sleeping; 3 → sleeping; . . . ; 10 → reading; etc.

   This is a function. Is it one-to-one?

490. Express $\frac{3}{8}$ as a percent.

540. When Lucy was getting only 8 hours of sleep each night, she could produce x ideas per hour during her 3 planning hours each day.

   When she was getting 10 hours of sleep, she could produce x + 4 ideas per hour during her 2 planning hours each day.

   Under the old system (8 hours sleep): x ideas/hour in 3 hours.

   Under the new system (10 hours sleep): x + 4 ideas/hour in 2 hours.

   Both systems produced the same number of ideas each day. Lucy was delighted. How many ideas per hour did she get under the old system?

   (Translation: Find x.) Use six pretty boxes.

660. How many ideas per hour under the new system?

785. Five-sixths of her ideas were worth nothing. One-sixth of them were worth $1,000 each. Under her new system, how much was she making per hour during her planning hours?

840. Using the results of the previous problem, what would have been Lucy's opportunity cost if she had continued doing the janitorial work, which she could have done for her at $18 per hour?

Chapter Twenty-seven
Most Important Concept in Economics

First part: Problems from this chapter

Some people like English better than math. Can you imagine that? These next three questions offer you the opportunity to write sentences.

129. Why would Prof. Eldwood never write a book entitled *A List of the Happiness Points Associated with Some Everyday Activities*?

260. When Lucy got her first train station, she was really happy. She might have assigned 1400 **HaPPiNeSS PoiNtS** to that event. When she gets her 485th train station, do you imagine that she will assign 1400 **HaPPiNeSS PoiNtS** to that?

304. Imagine working 50 feet off the ground on high-voltage electrical lines. That would scare me silly. What would happen if everyone assigned 0.0002 **HaPPiNeSS PoiNtS** to that work?

475. Some people assign a lot of **HaPPiNeSS PoiNtS** to creating new things. Lucy is most alive when she is working on her dream of creating the Lucy Railroad. She assigns 1,000 **HaPPiNeSS PoiNtS** to the hours she spends on that project. Joe might get 2% of that amount if he were thinking about creating a railroad. How many **HaPPiNeSS PoiNtS** is that?

600. Joe liked to eat popcorn while he watched fishing shows on television. One commercial fascinated Joe: The most delicious popcorn on earth! Phone in! Order any amount you want. We will deliver it to your door. All hot and buttery. Call now. Our operators are standing by.

 Joe didn't want to disappoint all those operators. He ordered 5 pounds of popcorn. In terms of pickup truckloads, how much was that? (One ounce of popcorn = 2,000 cubic inches. One pound = 16 ounces. One cubic foot = 1728 cubic inches. One cubic yard = 27 cubic feet. A pickup truck can hold 3 cubic yards.)

Second part: the 𝔐ixed 𝔅ag: a variety of problems from this chapter and previous chapters

731. Lucy had created a beautiful train station. She had entered a dangerous *Comfort Zone* in life. In his book *Roadblocks to Success,* 1844, Prof. Eldwood had warned his readers about this stage in life.

    Lucy was starting to lose her drive. Each morning it took her 3% longer to comb her hair than the previous morning. One June 1ˢᵗ it took her a minute to run a comb through her hair. How long would it be before she was spending two minutes combing her hair?

784. The title of the first chapter of *Roadblocks to Success* is NAH AGH. When the book was published in 1844, a lot of people went around saying, "NAH AGH." They hadn't read the book. They didn't know what that meant, but it sounded cool.

    ✶ He asked her to marry him. She would say, "NAH AGH."

    ✶ A mom would tell her son to milk the cows. He would respond, "NAH AGH."

    ✶ The weather reporter would announce on television, "Today's weather is NAH AGH."

    Weeks had gone by and Lucy had entered a *Comfort Zone* in her life. She was getting the habit of spending more and more time on her hair. She would wander around the train station admiring her work.

    Suddenly, she realized NAH AGH, which everyone nowadays knows means **N**ot **A**ll **H**abits **A**re **G**ood **H**abits.

    She listed all her habits and called that set H. She assigned each habit of hers into one of three categories: ① Those habits that she definitely wanted to keep; ② Those habits that neither helped nor harmed her; ③ Those habits she wanted to get rid of.

    She had 26 habits in her list (set H). Is this a function, and, secondly, is it one-to-one?

852. Name a number that stays the same when you square it.

910. Name a number that gets smaller when you square it.

950. When you square 7 you get 49. $7^2 = 49$
     What is the square of 6⅔?

Chapter Twenty-eight
Demand Curve

First part: Problems from this chapter

107. It really surprised Lucy. People were starting to wander into her train station. The building was attractive. The sign—**I LOVE YOU AMY STATION**—didn't really say that it was a train station. Everyone in town knew that those railroad tracks hadn't been used in years.

Some of them used the vending machines that Lucy had installed. Several of them asked the receptionist what this building was. On the receptionist's desk was a scrap of paper from Lucy's doll house days.

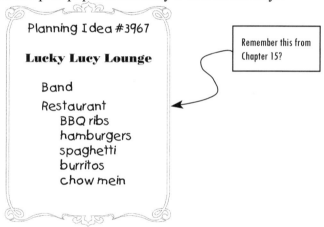

Planning Idea #3967

Lucky Lucy Lounge

Band
Restaurant
  BBQ ribs
  hamburgers
  spaghetti
  burritos
  chow mein

Remember this from Chapter 15?

The rumor spread.

Phone calls poured into the station. The receptionist could only say, "I cannot either confirm nor deny."

When politicians say these magic words, you can bet that it's probably true.

When the reporters came to interview Lucy, she was busy plotting the future of the Lucy Railroad. On her paper she had written: *If I have 2 trains, I'll need 1 station. With 3 trains I'll need 3 stations. With 5 trains I'll need 7 stations.*

Plot (2, 1), (3, 3), and (5, 7).

305. Newspapers love to report rumors. It became front-page news for the university paper.

THE KITTEN Caboodle

The Official Campus Newspaper of KITTENS University   Wednesday 3:10 p.m. Edition   10¢

exclusive

New Joint Opens!

artist conception

KANSAS: Everyone on campus is talking about the I Love Amy Station that is due to open this next Monday. Don't you just love the name! It is located at the site of the old Bedford train station that closed years ago.

It will offer the second largest menu in town—BBQ ribs, hamburgers, spaghetti, burritos, and chow mein. Only the 120 pizzas offered by Stanthony at his famous PieOne pizza place offers more.

Our KITTEN Caboodle reporter asked the hostess about whether reservations were being taken for the Grand Opening. She replied, "Reservations? I haven't heard anything about that." Obviously, it will be important to arrive early if you want a seat.

When Lucy saw the newspaper on Wednesday she laughed. She knew that newspapers often get their stories wrong. If Wednesday is day 1, then Lucy's panic level was zero.

On Thursday (day 2), people were calling the train station to ask for sample menus. Her panic level went up to 2.

On Friday (day 3), her receptionist asked Lucy what she should wear for Monday's opening. Lucy shouted, "What opening? This whole thing is a mistake." Panic level 5.

On Saturday (day 4), Lucy's mom asked her, "Where are you going to put the kitchen? How do you want the chairs arranged? Do you want Luke to help with the cleaning?" Lucy went to panic level 9.

You give me the day number and we have Lucy's panic level. It is a function from the day number to the panic level. $1 \rightarrow 0; 2 \rightarrow 2; 3 \rightarrow 5; 4 \rightarrow 9$. Plot (1, 0), (2, 2), (3, 5), and (4, 9).

351. Plot $y = x^3$ for values of x between 0 and 3.

Second part: the 𝕸ixed 𝕭ag: a variety of problems from this chapter and previous chapters

400. In Sunday school Lucy learned about how some bad guys threw Daniel into the fiery furnace and how God let him keep his cool. Lucy thought This is worse than a fiery furnace. My train station is turning into Lucky Lucy's Lounge tomorrow!

   That afternoon she called **KAMMY'S KWICK KITCHENS** and asked them to install a kitchen tomorrow morning at the I Love You Amy Station.
   She called her receptionist and asked her to hire some cooks and to print up some menus.
   She asked her mom to hire some servers.
   She asked her brother Luke to arrange for some music.
   If the domain = {Kammy's Kwick Kitchens, receptionist, mom, Luke} and the codomain is all the things that needed to be done, then this is *not* a function. Why not?

583. Solve $6x - 13 = 2x + 9$

632. After Lucy had made all the assignments, she settled into her inner office at the train station. Her regular outer office had the usual desk, chair, bookcases, etc. Her inner office was like the inside of a caboose. It was a place she could retreat into and do her deep thinking.

   Her thoughts: a kitchen, cooks, menus, servers, music. Have I forgotten anything? Then it hit her. Food!
   Out of her inner office, through her outer office, past the receptionist, through the train station, down the street, and into the grocery store.
   The first thing she spotted was cans of Alfredo brand chow mein. One can serves 2. Three cans for 49¢. Lucy needed to be able to serve 100 people. Using conversion factors, how much is this going to cost Lucy?

Chapter Twenty-nine
Whom Should You Trade With?

First part: Problems from this chapter

151. Did you ever notice that virtually everything government does is done using force. It taxes you (force) and then spends the money on police (force), the military (force), the courts (force), and zillions of regulations and laws (backed up by the threat of force).

Lucy was only 7 and didn't realize that it was against the law for her to hire her older brother—he was too young. It was against the law to open her train station for business unless she paid the local government for a piece of paper called a business license. It was against the law for her to open a can of chow mein, heat it, and serve it to KITTENS students unless she paid more money to the local government for another piece of paper called a restaurant license and submitted her place to a variety of government inspections. It was against the law for her to hire cooks unless she verified that they were legally in this country.

She didn't know about these laws, but when the government finds out about her being a crook, they are going to fine her hard and close her business.

Lucy wanted to buy a **Pammy Piano** for her new restaurant. On the Internet she found that Coalback Music Sales (in Kansas) was offering them for $3,000. She also found that she could buy them directly from the manufacturer in China for $2,200. This price included the shipping costs but not the tariff tax on imported pianos. That tax was $2,000. She had to buy from Coalback. The government tax forced her to do that.

Who was hurt by that tax?

233. What do you suppose would happen if the government stopped doing health inspections on restaurants?

Second part: the 𝔐ixed 𝔅ag: a variety of problems from this chapter and previous chapters

308. Lucy looked at the cans of spaghetti at Alfredo's Foods. A large can of Alfredo brand spaghetti cost 48¢ per can.

Alfredo brand spaghetti also comes in a smaller can. Sixteen of those smaller cans have the same amount as 12 of the bigger cans.

What would be the expected price of a smaller can?

493. Predicting the future is always a guess. That's true when you are investing in stocks, buying a home, or choosing a spouse. Lucy had to guess how many customers would be choosing ribs, spaghetti, and chow mein. She estimated 20 would choose ribs, 30, spaghetti, and 50, chow mein. Draw a bar graph illustrating this.

661. Lucy learned in Prof. Eldwood's *Easy Guide to Restaurant Pricing,* 1852, that the cost of the food is generally about 30% of the price on the menu. (The rest goes to labor costs, taxes, real estate costs, utilities, etc.)

If a serving of ribs would cost Lucy $2.88, what should the menu price be?

705. The division of labor makes a society more prosperous. If you look at societies where each family has to make their own clothes, grow their own food, build their own houses, you will usually find that their standard of living is low.

On the other hand, if each person can specialize and learn to do that one thing well, you often find prosperity. KAMMY'S KWIK KITCHENS arrived at 6 a.m. on Monday and were done in an hour.

It would have taken Lucy months to do what they did in minutes.

In economics the goal isn't to have everyone working 60 hours a week. It is to have abundant things.

In life is the goal to have abundant things?

841. Graph y = 4x − 3 for values of x from 1 to 5.

932. Draw a Venn diagram showing two sets: the set of everything that Lucy owns and the set of all the real estate in Kansas.

Chapter Thirty
David Ricardo's Law of Comparative Advantage

First part: Problems from this chapter

125. Back in Chapter 26 we learned why brain surgeons don't try to make money mowing other people's lawns. This is an illustration of Ricardo's Law of Comparative Advantage.

In one hour Lucy's mom can wash 200 dishes. Her son Luke can wash 40 dishes.

In one hour Lucy's mom can make 12 beds. Luke can make 4 beds.

Does Lucy's mom have an absolute advantage over her son?

306. Make the production chart.

356. In an hour Lucy's mom can either wash 200 dishes or make 12 beds. For each bed that she makes, how many dishes does she give up washing? (Translation: What is the opportunity cost for making a bed?)

441. In the previous problem, you did one of the four opportunity cost calculations. Do the other three. (Translation: Make the opportunity cost chart.)

494. Who should be doing dishes? (Translation: Who has the smaller opportunity cost?)

555. This is the production chart for Lucy, her receptionist, and Luke showing how much painting each can do (in cans of paint per hour).

Lucy	3
receptionist	2
Luke	1.5

The doorbell rings. Which one should go answer the door?

Second part: the 𝔐ixed 𝔅ag: a variety of problems from this chapter and previous chapters

662. This is an **opportunity chart** for two choices that Lucy has.

	Lucy Railroad	Lucky Lucy's Lounge
Lucy		$\frac{3}{5}$

The $\frac{3}{5}$ came from a **production chart** in which the railroad could make $3,000,000 in a year and the lounge could make $5,000,000 a year—or was it that the railroad could make $5,000,000 a year and the lounge make $3,000,000 a year?

740. Fill in the missing entry in that **opportunity chart**.

911. If the receptionist's salary is currently $18/hour and it is increased by 6% each year, how much will her salary be in one year and in 10 years?
    For the 10 years, you may leave your answer in the form ■ × (■)".

951. Fill in the missing entries in this **opportunity chart**.

	product A	product B
Lucy	$\frac{4}{5}$	
receptionist		$\frac{7}{3}$
Luke	1.5	

Chapter Thirty-one
Why Ricardo Is Right

First part: Problems from this chapter

153. The central part of the proof of Ricardo's Law of Comparative Advantage was showing that if you invert two fractions, the "sense" of the inequality changes.

        Translation: if $a/b < c/d$, then $b/a > d/c$.

        Let's try it with a pair of fractions, say, $\frac{2}{3}$ and $\frac{3}{4}$

        First of all, which of these two is smaller?

257. Then if you invert $\frac{2}{3}$ and $\frac{3}{4}$ which one is smaller?

355. Take two positive numbers, x and y, where $x < y$. Is it always true that $x^2 < y^2$?

479. Monday came. The kitchen had been installed. The food had been purchased. The prices on the menu had been determined—30% of menu price = price of food. The menus had been printed. The tables and chairs were in place. The cooks and servers were ready. Luke had brought some of his music CDs and a CD player.

        Lucy was tired, really tired. She had been awake for many hours having gotten everything ready for the grand opening of **Lucky Lucy's Lounge**. She headed to her inner office, which looked like the inside of a caboose . . . and fell asleep. She missed the first hours of her new business.

        The receptionist and Luke had to run the show. In an hour the receptionist could answer 8 phone calls or seat 11 parties. Luke could answer 5 phone calls or seat 7 parties. Who should do what?

Second part: the 𝔐ixed 𝔅ag: a variety of problems from this chapter and previous chapters

541. Lucy woke at 11 a.m. The receptionist had hired **Vanny's Vans** to haul the receipts off to the bank.

   One $5 bill fell off the van.
What percent of the $400,000 was that?

670. Earlier this morning Lucy had been paying her brother Luke $8/hour. She raised his salary to $11. What percent increase was that?

741. In the first hour of Lucky Lucy's Lounge, each order of burritos gave a profit of $21, and each order of chow mein gave a profit of $14. There were 30 more orders for chow mein than for burritos. The total profit in that first hour from these two menu items was $1,365.

   How many burrito orders were there? Use six pretty boxes.

842. Do the previous problem using "Let x = . . ." statements.

930. For every value of x that you can name, $y = \frac{x}{4}$ gives you a value of y. For example, if x is 40, then y is 10.

   If the domain and codomain are the natural numbers {1, 2, 3, . . .}, is the assignment $y = \frac{x}{4}$ a function?

Chapter Thirty-two
What Economics Is All About

First part: Problems from this chapter

132. The smallest cardinal number is zero. Cardinal numbers count the number of elements in a set, and you can't have less than zero elements in a set.

What is the smallest ordinal number?

207. Lucy called Alfredo Foods and asked them to deliver some cans of Alfredo brand chow mein to Lucky Lucy's Lounge.

"How many would you like?"

Lucy said, "A truckload would be fine."

"That will cost you $2,000."

Lucy asked, "Do you give any discount for cash?"

"Sure. Cash on delivery and I'll give you a 14% discount."

How much money should Lucy have for that delivery?

319. By noon 1,900 people had dined at Lucky Lucy's Lounge. Some of them had paid in cash. Some of them had ordered ribs. Some of them had done both: they ordered ribs and paid in cash. Draw a Venn diagram.

403. Here is how we might use that Venn diagram that you just drew. Suppose 700 had paid in cash and 400 had ordered ribs. Suppose 100 had done both.

How many had paid in cash but didn't order ribs?

584. How many ordered ribs but didn't pay in cash?

633. How many of those 1,900 people didn't order ribs or pay in cash?

Second part: the 𝔐ixed 𝔅ag: a variety of problems from this chapter and previous chapters

706. The sixth point of economics is: *If you want a successful business, you need to find a significant need that others don't see.*

Lucy had read that on page 284 of *Life of Fred: Pre-Algebra 2 with Economics.*

By 1 p.m. **Vanny's Vans** had hauled off a second truckload of cash to the bank. Lucy headed back to her inner office that looked like a caboose and went into her thinking and planning mode.

She wrote on a piece of paper:

Fact #1: Those railroad tracks haven't been used in six years.

Fact #2: My railroad station had so little value as a train station that my wonderful grandma gave it to me.

Fact #3: There is a ~~big~~ huge need for what we have offered today.

Lucy had crossed out "big."
Her father had told Lucy, "Opportunity knocks only once."
Her mother had told her, "Opportunity sometimes only scratches lightly at the door."
Lucy knew that opportunity was banging hard on her door.

Lucy's brother said, "You are so lucky."
Lucy knew that Luke spent his time playing tag with his friends while she had been reading and thinking about the business world for years.
She called **Rammy's Real Estate** and asked Rammy to locate all the vacant train stations in Kansas.
She called **Wammy's Westeraunt Supplies** and spoke with Wamford. She told him that she would be needing lots of tables, chairs, knives, forks, spoons, cups, and napkins. Wamford was delighted.
She called **Hammy's Human Resources** and spoke with Hamlock about getting cooks and servers for restaurants she would be opening.

Your question: Lucy could have done the work herself that she was hiring Rammy, Wamford, and Hamlock to do. Why didn't she?

800. Rammy located 96 vacant train stations in Kansas.
        He reported to Lucy:
34 have air conditioning installed
35 have good roofs
43 have unbroken windows
        7 have air conditioning and good roofs
        10 have air conditioning and unbroken windows
        9 have good roofs and unbroken windows
                3 have air conditioning, good roofs and good windows

How many have no air conditioning, bad roofs, and broken windows?

851.   Rammy: I assume you want the 3 nice ones with air conditioning, good roofs, and good windows.
        Lucy: Silly man! I want all 96 of them.
        Rammy: But anyone who knows about Venn diagrams can figure out that 7 of them are really bad. They have no air conditioning, bad roofs, and broken windows.
        Lucy: Those crummy ones will cost less. After the costs of fixing them up, they will cost the same as the 3 perfect ones.
        Rammy was trying to sell Lucy 3 stations. She wanted to buy all 96 of them. To the nearest percent how much more is 96 than 3?

_____

Note to readers: Venn diagrams will be mentioned in beginning algebra. Venn diagrams with numbers in them is an advanced topic. It will be first introduced in *Life of Fred: Advanced Algebra Expanded Edition.* Only one problem in that book will have the very advanced "three circles" (like the one you did in #800 above).

page 115, #5:   There are 100 cars in the parking lot.
                27 of them are red.
                41 are Fords.
                12 are convertibles.
                6 of them are red Fords.
                9 of them are Ford convertibles.
                5 of them are red convertibles.
                4 of them are red Ford convertibles.
                How many of them are not red or Fords or convertibles?
        When you did #800 above, I didn't want to scare you and tell you that it was a hard advanced algebra topic. (By the way, the answer to #5 is 36 cars that are not red or Fords or convertibles.)

Chapter Thirty-three
Dividing the Pudding

First part: Problems from this chapter

152. Dividing the pudding into six equal shares so that no one can complain—that's the central hard problem of this chapter.

There's an easier problem, which we are going to tackle right now. Suppose you have something that can't be divided. In Aunt Mabel's will suppose she left you and your five siblings* her waffle iron. There is no way that you can chop it into six pieces. Several of you have fond memories of Aunt Mabel making waffles for you on Saturday mornings when you would visit her.

It's is virtually worthless to other people. So selling it and splitting the proceeds won't work. And, besides, it wouldn't be worth the trouble to clean it up, advertise, and sell it.

Make a suggestion how you and your siblings might deal with this inheritance.

236. Let's assume you have read my answer to the previous question. Suppose Aunt Mabel's will left 40% of the waffle iron to you and 12% to each of your other siblings.

How do you handle the sealed bids?

357. The Lucky Lucy Lounges will be worth millions in a couple of years. Lucy is planning on getting married when she gets older, and she hopes to have six kids. When she dies, she wants to leave her business to her kids. Businesses are a little like waffle irons—they can't be cut up into six parts. Sealed bids by the six kids are a lot better than fistfights. Lucy will put a "sealed bids" clause in her will.

But there is one drawback. With Aunt Mabel's waffle iron, all the nephews and nieces each had some spare money to bid with.

None of Lucy's kids have the millions it would take to make a bid. What else might Lucy put in her will so that her kids can make bids?

――――――――――――――――

* A sibling is a brother or sister.

Second part: the 𝔐ixed 𝔅ag: a variety of problems from this chapter and previous chapters

480. When Lucy was younger and had little money, she did all the work of fixing up the first station by herself—painting, electrical, nailing, and so on.[*]

　　With the 96 vacant train stations in Kansas that Lucy will be buying, she will be using the **division of labor** concept from economics. More things get done if each one does their specialty, if each one pays attention to their opportunity costs.

　　When Lucy plans and negotiates, she makes (on the average) $2,000 per hour. If she paints walls, she makes $37 per hour. If she sweeps floors, she makes $9 per hour.

　　What is her opportunity cost if she paints walls?

557. Before Lucy's 10 a.m. nap, which she takes almost every day, she wrote emails answering customer questions, getting bids on electrical work, asking for advice on taxes, etc. She found that after her morning nap she could write 7 more emails per hour than before her nap.

　　In the 2 hours before her nap and the 2 hours after her nap, she wrote a total of 158 emails. What was her rate of writing emails before her nap?

672. Solve $6.2x + 17 = 2.4x + 22.7$

　　If it took Lucy 6 days to renovate each of the 96 vacant train stations she bought, you might wonder how long it would take her to do all of the stations. The answer is 6 days. The division of labor means that she is hiring others. All 96 stations are being worked on at the same time.

---

[*] Parallel constructions is one thing my English teachers emphasized. The parts should match each other. You don't write: *Lucy was successful because of her work and her being dedicated.* Instead: *Lucy was successful because of her work and her dedication.*

　　I wrote: "... painting, electrical, nailing. ..." I can't think of how to write: *painting, electricaling, nailing.* My English teachers would probably be disappointed in me.

First part: Problems from this chapter

126. In less than a fortnight* there were 97 I Love You Amy Stations in Kansas.

Intermission

I mentioned Middle English in the footnote. In a recent survey 97% of my readers don't know what that is.

One reader said, "If ABCDEFGHIJKLMNOPQRSTUVWXYZ is English, then GHIJKLMNOP is probably Middle English."

No.

Middle English was written and spoken 500+ years ago. (1150–1500 to be more exact) English is not like math. It changes a lot.

Geofferey Chaucer wrote *Canterbury Tales* using Middle English (because he lived between 1150 and 1500). The first four lines begin:

Whan that aprill with his shoures soote
The droghte of march hath perced to the roote,
And bathed every veyne in swich licour
Of which vertu engendred is the flour. . . .

This is English. And all the words are spelled correctly (according to Middle English). Our Modern English will be just as unclear to readers 500 years from now as this Chaucer stuff is to us today.

The word "today" was spelled "to-day" in the 1800s. In about 1900 "today" first started being used. By the 1920s it became more common. Jack London's stories are filled with "to-day."

In case you are curious . . .

When April with his showers sweet with fruit
The drought of March has pierced unto the root
And bathed each vein with liquor that has power
To generate therein and sire the flower. . . .

---

* A fortnight is 14 days. In Middle English it was *fourtenight,* which was contracted down to *fortnight* in Modern English.

Intermission
continued

Can you believe it? Some people still spell email as e-mail. It's the same story as today and to-day.

Where were we? Oh yes. Lucy had created 97 wonderful places to eat in Kansas. Here is a map of Kansas, which shows the (approximately) 47 cities where all the I Love You Amy Stations are located.

If we create a function whose domain is the 97 I Love You Amy Stations and the codomain is these 47 cities in which these stations are all located (and where the rule is the obvious one—map each station to the city it's located in) will this function be one-to-one?

261. $y = 10x^2 + 7$
The variable x is measured in days from when Lucy asked Rammy to locate the vacant railroad stations. The variable y is Lucy's net worth in millions of dollars. Plot this curve from $x = 0$ to $x = 6$.

320. Try out different values of x (from the previous problem) and find out when Lucy will be a billionaire. Translation: When she will be worth more than $1,000,000,000.

Second part: the 𝔐ixed 𝔅ag: a variety of problems from this chapter and previous chapters

440. Lucy was 7. There was one part of childhood that she had missed. Unlike when she was younger, she now wanted a doll. A regular, cute, loveable doll. Luke had had a Rag-A-Fluffy doll when he was little. Those dolls had been very popular and it seemed that everyone had one. She wanted something special. A cute little boy doll—one that supposedly knew a lot about mathematics.

genuine Fred Gauss doll

After Lucy got one, she learned that her father also had one, which he kept on his desk at **Tammy's Tires**. He said that the little fellow seemed to be a comfort when the stresses on his job got to be too much.

Is the number of Fred Gauss dolls a discrete variable or a continuous variable?

624. Suppose some shoe is manufactured in sizes 8, 8½, 9, 9½, 10, 10½, 11, 11½, and 12. Is this discrete or continuous?

671. Luke had ordered a Fred doll. He called it an "action figure" so that his friends wouldn't make fun of him. Under Luke's direction, Fred would go on all kinds of adventures. Luke would give Fred a hug 4 times each hour to give him encouragement.

Lucy's dad would give his Fred doll a hug 5 times each hour. He called it a "good luck squeeze" so that his fellow workers wouldn't make fun of him.

Lucy hugged her Fred doll 6 times each hour. All her girl friends were jealous that she had such a lovable doll and they didn't.

How long would it take these three people to give a total of 50 hugs?

853. In the first month that Luke got his Fred doll, he sent Fred off on 70 adventures. Each month thereafter Luke would send Fred off on 18% more adventures than the previous month. How long would it be before Fred was going on 140 adventures each month?

933. Solve $5(x + 2) + 12 = 39$

104. Definition: A **function** is any rule that associates each element of the first set (the domain) with exactly one element of the second set (the codomain).

Suppose the first set (the domain) is the set of all teachers at KITTENS University. Suppose the second set (the codomain) is the set of all numbers.

Is this a function? *Associate to each teacher the salary that they are now making each month.*

Yes. By this function Fred is assigned to $500. 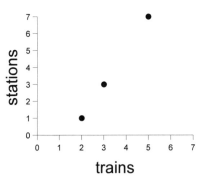 is assigned to $6,700. Each teacher is assigned to exactly one number.

106. The person with the highest opportunity cost is the . . .

A) most likely

B) least likely

person to take a job working at **Tammy's Tires**?

Suppose **Tammy's Tires** pays $24/hour. If you were making $30/hour, would you take a job at **Tammy's Tires**? No. Your opportunity cost is $30/hour.

If your friend Mark is making only $11/hour working at **Cammy's Custom-Cut Carpets**, he would be happy to switch jobs. Mark's opportunity cost is $11/hour. That is what he would be giving up.

The person with the highest opportunity cost is the A) least likely to take a job working at **Tammy's Tires**.

107. If I have 2 trains, I'll need 1 station. With 3 trains I'll need 3 stations. With 5 trains I'll need 7 stations.

Plot (2, 1), (3, 3), and (5, 7).

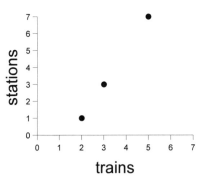

109. Lucy was spending ten hours each day immersed in all aspects of rail road work. Roughly, how long will it take her to amass 5,000 hours of learning?

$$\frac{5{,}000 \text{ hours of learning}}{1} \times \frac{1 \text{ day}}{10 \text{ hours of learning}}$$

$$= \frac{5{,}000 \text{ hours of learning}}{1} \times \frac{1 \text{ day}}{10 \text{ hours of learning}} = 500 \text{ days}$$

Lucy was really dedicated to learning about trains. Many people would think that they were really dedicated if they spent one hour a day, seven days a week, learning about the business they were thinking about starting. Of course, this would take ten times as long—comparing Lucy's ten hours a day with one hour per day. It would take 5,000 days to get ready. That's about 13.7 years.

If you are working at a full-time job, then one hour of learning is a lot—and 13.7 years is a long time.

Now you can see why the years before you turn 18 and head out into the working world are so important. Waste that time and you will have to be like most people—people who have no choice—people who will have to work for others for years and years and years.

Pick something . . . now! Something that excites you. It can be silly or it can be serious. Animals in the ocean (marine biology); raising horses; pizza restaurants; making shoes; oil painting; owning a circus; building houses; computers; writing books about cooking turkeys; industrial air conditioning; owning and running a hospital; apple orchard management; making watches; producing movies about Irish dancing. . . .

I, your reader, think that whole idea is funny. How am I suppose to choose something—say creating machines that wash ducks—at the age of 4 and expect to stick with that for the rest of my life?

You probably won't be stuck with machines that wash ducks for the rest of your life. It will probably change a bit over the years. After you study about the machinery that might do that job and after you study all about ducks, you might drift into machines that wash pigs instead—especially, after you find out that ducks will wash themselves if you stick them in a lake. Then you might drift into designing computer programs that work with machines that wash pigs. Then it might be computer programs that work with machines that turn pigs into pre-cooked bacon. And so on.

Learn from Lucy. Pick something and start focusing.

114. She kept a diary of all the things she learned about railroading. Each week she wrote r ideas down about the costs of buying railroad things—cars, stations, land. . . .

Each week she wrote down $r + 5$ ideas about what her customers would like to have in dealing with Lucy Railroad.

After 20 weeks Lucy had a total of 580 ideas in her diary. How many ideas per week did she write down regarding buying railroad things?

	d total ideas	r ideas per week	t weeks
ideas for buying		r	
ideas for customers' needs			

	d total ideas	r ideas per week	t weeks
ideas for buying		r	20
ideas for customers' needs		$r + 5$	20

	total ideas	ideas per week	weeks
ideas for buying	$20r$	r	20
ideas for customers' needs	$20(r + 5)$	$r + 5$	20

We know that she wrote down a total of 580 ideas.

$$20r + 20(r + 5) = 580$$

Distributive law $\qquad 20r + 20r + 100 = 580$

Combine like terms $\qquad 40r + 100 = 580$

Subtract 100 from both sides $\qquad 40r = 480$

Divide both sides by 40 $\qquad r = 12$

Lucy had 12 ideas per week regarding buying railroad things.

118. Prof. Eldwood wrote: In the event it happens that two or more trains are situated on the same track and heading towards each other at a high rate of speed, it should be noted that it is essential that in the interests of good railroad management that steps be taken to avoid the situation in which impaction occurs. Clean up that sentence as much as you can.

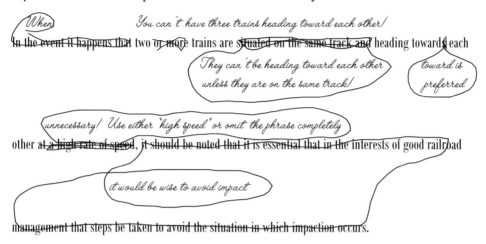

When

You can't have three trains heading toward each other!

~~In the event it happens that~~ two ~~or more~~ trains are ~~situated on the same track and~~ heading towards each

They can't be heading toward each other unless they are on the same track!

toward is preferred

unnecessary! Use either "high speed" or omit the phrase completely

other ~~at a high rate of speed, it should be noted that it is essential that in the~~ interests of good railroad

it would be wise to avoid impact

~~management that steps be taken to avoid the situation in which impaction occurs.~~

Or how about just saying: Trains shouldn't collide.

You don't even have to mention "two trains" since it's hard to have just one train collide. You don't have to mention that they are heading toward each other, because that's the only way that trains can collide.*

119. Are there numbers that are not on the number line?
Translation: That means are there numbers that are not less than zero, equal to zero, or greater than zero?

What if there were a number that was right here?

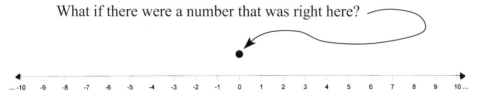

Suppose there were a number sitting *above* the number line.

(continued on next page)

---

* I take that back. One train could run into the rear end of the other train. But saying, "Trains shouldn't collide" covers even more cases than Eldwood's trains heading in opposite directions.

101

Impossible! I, your reader, can't be fooled. Every number has got to be on the number line. π, 9967.3, –30, ⅞ are all on the number line. I double-dog dare you to name a number that is not less than zero, not equal to zero, and not greater than zero.

i

That's it? i You got to be kidding.

Would I kid you? The number i sits one unit above zero on the number line. It's right where I drew that black dot. ●

Right now, you are in pre-algebra. Then comes beginning algebra (also known as algebra 1). In beginning algebra we will tell you that every number is either less than zero, equal to zero, or greater than zero. It is called the **Law of Trichotomy**.

Then comes advanced algebra. We will first tell you that the Law of Trichotomy is true *for every number on the number line*. Then we will show you i.

I can hardly wait.

Read faster. ☺

120. Lucy had Luke's engine and 53 train cars. The engine, which had a motor in it, cost \$3.39. The other non-engine cars each cost x dollars. Luke had bought the whole collection for \$129. How much did each of the other cars cost?

Let x = the cost of one of the non-engine cars.

Then 53x = the cost of all the non-engine cars.

Then 53x + 3.39 = the cost of the whole collection.

At this point we can write the equation.

$$53x + 3.39 = 129$$

Subtract 3.39 from both sides $\quad\quad 53x = 125.61$

Divide both sides by 53 $\quad\quad\quad\quad x = 2.37$

Each of the non-engine cars cost \$2.37

121. Which of these pairs is commutative?

A) Owning a train station

B) Painting that station........................No. If you paint someone else's train station and it isn't yours, the owner may call the cops.

A) Getting off the train

B) Going inside the station............No. It would be really hard to go inside the station first and then get off the train.

A) Using a vending machine in the station

B) Using the restroom in the station..............Yes. You can do these in either order and the result will be the same. I'm excluding the case where you really have to go to the bathroom and stopping first to buy some gum from a vending machine could result in a disaster.

A) Owning a train station

B) Selling that train station............No. It is tough to sell what you don't own. If you sell someone a train station for $195,000 and you don't own it, the buyer will be really unhappy. As my younger daughter said when she was little, "You could get killed to death."

122. Let's look at it in reverse. Suppose it cost $1,000 to remove those poles. How much should each family member pay?

　　Since they owned the front lawn in common and since each one would receive the same benefit, it would only seem fair that each should bear 25% of the cost. Each should pay $250.

　　Wait a minute! I, your reader, object. Lucy stuck those stupid poles in the front lawn. She should pay the whole $1,000.

　　That's the whole point of why it is called a *tragedy*. The unfairness came when she used that common property for her own personal gain. It was at that point she should have compensated the other three owners.

123. It would take 5 days to get the whole job done. What fraction of the work would be done each day?

　　One-fifth of the job would be done each day.

125. In one hour Lucy's mom can wash 200 dishes. Her son Luke can wash 40 dishes. In one hour Lucy's mom can make 12 beds. Luke can make 4 beds.

Does Lucy's mom, have an absolute advantage over her son?

Yes. She can wash more dishes and she can make more beds.

126. If we create a function whose domain is the 97 I Love You Amy Stations and the codomain is these 47 cities in which these station are all located (and where the rule is the obvious one—map each station to the city it's located in—will this function be one-to-one?

It can't be 1–1. At least two of the stations will have to be assigned to the same city.

129. Why would Prof. Eldwood never write a book entitled *A List of the Happiness Points Associated with Some Everyday Activities*?

What makes you super happy may not be the same as what makes me super happy. The assignment of **HaPPiNeSS PoiNtS** is a very individual matter.

I assign 1000 **HaPPiNeSS PoiNtS** to getting up every morning and writing the *Life of Fred* books from 4 a.m. to 6 a.m. I really like doing that. I sometimes giggle when I write about Lucy. On the other hand, many people (99.9999% of the world) would assign about 0 **HaPPiNeSS PoiNtS** to that activity.

Prof. Eldwood can't tell me what would make me happy.

132. The smallest cardinal number is zero. Cardinal numbers count the number of elements in a set, and you can't have less than zero elements in a set. What is the smallest ordinal number?

The ordinal numbers are *first, second, third, fourth,* and so on. The smallest ordinal number is *first.*

If you are first in line, there can't be anyone ahead of you.

136. The Pythagorean theorem says
   that in any right triangle ☞

A right triangle is a triangle with a right angle in it.
A right angle is the angle in a corner of a square.

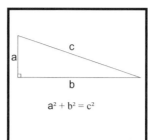

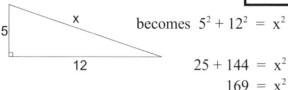 becomes $5^2 + 12^2 = x^2$

$$25 + 144 = x^2$$
$$169 = x^2$$

At this point we might be stuck.

The equation $169 = x^2$ asks, "What number times itself equals 169?"
We can try out different numbers.

If x = 10, then $169 = x^2$ becomes $169 \stackrel{?}{=} 100$ No.

If x = 11, then $169 = x^2$ becomes $169 \stackrel{?}{=} 121$ No.

If x = 12, then $169 = x^2$ becomes $169 \stackrel{?}{=} 144$ No.

If x = 13, then $169 = x^2$ becomes $169 \stackrel{?}{=} 169$ Yes!

The distance from Rose to Dawn is 13 miles.

Wait a minute! I, your reader, have a small question. It is super easy to solve equations like $x^2 = 49$. Anybody can see that the answer is 7. But suppose the problem had been

Then you would get $2^2 + 5^2 = x^2$.
This is $4 + 25 = x^2$.
What are you going to do with $29 = x^2$?
Ha! Ha! Ha! Ha! Ha! Ha! Ha! Ha! Ha! Ha! Ha! Ha! Ha! Ha! Ha! Ha!

There is no exact decimal answer. When you get to beginning algebra, you will learn about square roots. The square root of 49 is 7. In symbols, $\sqrt{49} = 7$. The square root of 64 is 8. In symbols, $\sqrt{64} = 8$. The square root of 100 is 10. In symbols, $\sqrt{100} = 10$.

On your calculator is a $\sqrt{}$ key. Play with it. Type in 144 and hit that key and you get 12.

You ask me for an approximate answer for $\sqrt{29}$? On my big calculator I get $\sqrt{29} \approx 5.38516480713450403125071049915403$.

144. She ran to her bedroom and gathered up some of her other yucky dolls. She was going to trade 4 dolls for every 3 train cars. How many dolls would she need to trade for Luke's 51 train cars?

There are two ways to do this problem. Both work equally well.

Using a proportion: 4 dolls match up with 3 train cars in the same way that x dolls match up with 51 train cars.

$$\frac{4 \text{ dolls}}{3 \text{ train cars}} = \frac{x \text{ dolls}}{51 \text{ train cars}}$$

$$\frac{4}{3} = \frac{x}{51}$$

cross multiply* $(4)(51) = 3x$

do the arithmetic $204 = 3x$

divide both sides by 3 $68 = x$

Lucy could trade 68 dolls for 51 train cars.

Using a conversion factor: Since 4 dolls match up with 3 train cars, the conversion factor will be either

$$\frac{4 \text{ dolls}}{3 \text{ train cars}} \quad \text{or} \quad \frac{3 \text{ train cars}}{4 \text{ dolls}}$$

We want to convert 51 train cars into dolls and choose the conversion factor so that the train cars cancel.

$$\frac{51 \text{ train cars}}{1} \times \frac{4 \text{ dolls}}{3 \text{ train cars}}$$

$$= \frac{(51)(4)}{3} \text{ dolls}$$

$$= 68 \text{ dolls}$$

---

* Cross multiplying was explained on three pages in *Life of Fred Pre-Algebra 1 with Biology*. See the index of that book for those pages.

    In brief . . . If you have $\frac{a}{b} = \frac{c}{d}$, then it will be true that $ad = bc$.

145. What would be the drawbacks for Lucy if she really did put Lucky Lucy's Lounge with a band playing quiet music, and a restaurant featuring BBQ ribs, hamburgers, spaghetti, corn on the cob, burritos, chow mein, and six flavors of Sluice, a three-screen movie theater and a skating rink and a shoe store **in her train station?**

This is a problem of the division of labor. It is so important that Lucy stay *focused* on creating the Lucy Railroad. She only has 24 hours in a day to think about and work on her railroad. For many people it is so easy to get *distracted* away from the main goal(s) of their life.

If you want to do railroads, then you don't want to concentrate on getting a band, which involves the size of the band, the government permits, the type of music they would play, any union requirements, whether to hire individuals (and issue paychecks and file government forms), etc.

You don't want to have a restaurant with a large menu and deal with government health inspections, hiring personnel, setting up a kitchen, etc.

Movie theaters, skating rinks, and shoe stores would each offer their own headaches.

Maybe later, after Lucy has her railroad in full operation, she might hire people to manage all these fancy additions to her train stations. Hire them, not do them herself.

Even opening a restaurant requires its own 5,000 hours of specialized education. There are not enough hours in 80 years of life to do everything.

Even a hundred lifetimes are too short to do everything. *Focus!*

146. One way to stop everyone from becoming rich is to pass a law that restricts poor, uneducated people from trading their labor for money. Luke would probably be delighted to work for $9/hour sweeping the floors in Lucy's train station. Lucy would be delighted to pay him $9/hour for his work, but if the government has a minimum wage law that says that employers must pay employees at least $10/hour, neither Lucy nor Luke are happy.

Who does benefit from that minimum wage law?

All those who are bigger and stronger who can do work worth at least $10/hour. They are glad to get rid of the competition that Luke offers. It is the advantaged that benefit from a minimum wage law. The less abled are discriminated against.

151. On the Internet she found that Coalback Music Sales (in Kansas) was offering **Pammy Pianos** for $3,000. She also found that she could buy them directly from the manufacturer in China for $2,200. This price included the shipping costs but not the tariff tax on imported pianos. That tax was $2,000. She had to buy from Coalback. The government tax forced her to do that.

Who was hurt by that tax?

It certainly wasn't Coalback. He loved the fact that the government forced Lucy to buy from him.

It certainly wasn't the government. From the few people who bought accidentally bought from the manufacturer in China, the U.S. government collected money from their tariff tax.

Lucy got hurt from this tariff law. She had to pay more because of the law.

Increasing Lucy's expenses meant that Lucy would have to charge more to stay in business. The general public who visited her restaurant would have to pay more.

152. In her will suppose Aunt Mabel left you and your five siblings her waffle iron. Make a suggestion how you and your siblings might deal with this inheritance.

Emails, cards, letters, telegrams, and phone calls have been pouring in with readers' suggestions. Here are some of them:

♣ Roll a die and the winner gets the waffle iron.
(Aunt Mabel wouldn't approve of gambling. She didn't even like playing cards.)

♣ Have a big fistfight and the winner gets the waffle iron.
(Aunt Mabel used to like watching wrestling on television, but having her six nephews/nieces in a big, bloody fight is different.)

♣ Every Sunday the waffle iron gets passed to another one of the siblings. (This might work, but when your sister Misselthrop moves to Alabama and your brother Raggledink moves to Oregon, this would get to be a big hassle.)

♣ Hold an auction and the winner pays the other five siblings. (This has some promise. The only difficulty I can see is if emotions start to get in the way during the auction. If you sister Widdle is hated by your sister Alleshka, then Alleshka might keep bidding against Widdle just to make sure she doesn't get that waffle iron. It would be tragic to have bad feelings because of a waffle iron.)

Widdle: I bid 25¢.
Alleshka: I bid 26¢.

Widdle: I bid 50¢.
Alleshka: I bid 51¢
Widdle: I bid a buck.
Alleshka: I bid a buck and a penny.
Widdle: I bid $3.
Alleshka: I bid $3.01.
Widdle: I bid $10.
Alleshka: I bid $10.01.

At this point Widdle may want to do bodily harm to Alleshka.

I, your reader, can't think of any other suggestions. Do you, Mr. Author, have one?[*]

You would be really surprised if I didn't.

I do have one suggestion.

Each person gets a piece of paper and puts their name and the amount that they would be willing to pay for the waffle iron. Each paper is put in an envelope and they are all put on the table and opened. The winner pays the other five siblings.

This is sometimes called a "sealed bid" auction.

153. Which of these two is smaller? $\frac{2}{3}$ or $\frac{3}{4}$

To compare two fractions you *pretend* that you are going to add them. You make their denominators alike.

$$\frac{2}{3} = \frac{8}{12} \qquad \frac{3}{4} = \frac{9}{12} \qquad \text{Now it's easy to see that } \frac{2}{3} < \frac{3}{4}$$

---

[*] This is what is called a rhetorical question. (re-TORE-eh-cul) A rhetorical question is a question where you don't expect a reply. Asking "How are you?" is a rhetorical question.

160. Let D = the set of all 32 Adventure Duck comic books that Luke bought. Let T = the set of 6 train books that Lucy got from the KITTENS library.

We know that assigning every member of D to Prof. Eldwood's *Locomotive Electrical Diagrams* book is a function.

Your question: Is this function one-to-one?

If two elements of the domain have the same image, then it is not a one-to-one function. Since all 32 members of the domain have the same image, this is really truly, definitely is not a one-to-one function.

161. What percent increase was it in going from {yucky} to {yucky, locomotive, caboose}?

Lucy went from a one-word vocabulary to a three-word vocabulary.

She gained two new words.

Two new words is what percent of one?

2 = ?% of 1

If you don't know both sides of the *of*, you divide the number closest to the *of* into the other number.

$2 \div 1 = 2 = 200\%$

Her vocabulary had increased by 200%.

162. Suppose the government decided that we must use tall trees for money. Only trees over a hundred feet tall could be used. Why would this not be a good choice?

Durable? Yes. Tall trees last for many years.

Portable? No. Imagine taking some tall trees to the grocery store.

Limited supply. Yes. No one can easily and quickly produce a lot of 100-foot trees.

Fungible? Maybe. If the government declared that any two tall trees had the same value, then they would be fungible. But if people preferred redwood trees over pine trees, then all trees wouldn't have the same value.

Divisible? No. If it took six tall trees to buy a car, then there wouldn't be any tall-tree-money to buy bubble gum.

163. Lots of people like socialism. Why?

If you take the Lucy Railroad and give ownership to the group (or the government or the collective), then *who benefits*? It is the ones who didn't do the work of creating that railroad. They all get a share of what they didn't own before.

Lucy wouldn't be *sharing* her railroad. The word *sharing* means a *voluntary* act. When socialists use the word *sharing*, they are distorting the meaning of that word.

What do you call it when something is involuntarily taken? The word is STEALING.

The definition of a slave
is someone who does not own
the fruits of their own labor.

164. Beside collusion and lying about your competitors, what is the third way to eliminate competition?

Males, especially, are inclined to use force—or the threat of force—to get their way. It's just a fact of nature. Bullying happens . . .

✤ on the playground. The big kid takes the little kid's ball.

✤ in high school. "You stay away from her or I'll knock your teeth in."

✤ in business. The mafia would threaten to break your legs if you offer competition.

✤ between nations. "We'll will nuke your cities."

It's fraud (previous chapter) and force (this chapter) that we need government to protect us against.

165. Lucy Railroad began with $6,000, and suppose it grew at the rate of 2% per month. How much would it be at the end of one month?

This is a "2% more" problem.
We will do it the easy way.
2% plus the original 100% is 102%.
102% of $6,000
$1.02 \times 6{,}000 = \$6{,}120$ at the end of one month.

166. Lucy sent in $10,000. The statues were $500 each. There was a $1,000 shipping charge. How many statues would she get?

Let x = the number of statues that Lucy would receive
Then 500x = the cost of those statues
Then 500x + 1,000 = the total amount she paid

$$500x + 1,000 = 10,000$$

Subtract 1,000 from both sides $500x = 9,000$
Divide both sides by 500 $x = 18$

She would receive 18 statues.

Sharp readers might have noticed that this is the third time we have come up with this equation. In *Life of Fred Pre-Algebra 2 with Economics* we got this equation on pages 53–55 when Fred was heading toward Main Street. Then we got this same equation in the *Your Turn to Play* in problem #1 in working with your weekly allowance. Now we have it in Lucy's buying train station statues.

The important point is that the same old algebra can be used in a great variety of circumstances. One algebra = many uses.

167. Lucy would spend eight hours on railroads—reading about, talking about, thinking about, dreaming about trains on each day, Monday through Saturday. How many hours of "train time" would she accumulate in 52 weeks?

We want to convert 52 weeks into hours of train time.

$$\frac{52 \text{ weeks}}{1} \times \frac{48 \text{ hours of train time}}{1 \text{ week}} = 2,496 \text{ hours of train time}$$

In two years Lucy would have devoted almost 5,000 hours toward her life goal.

168.   They worked for a total of 5 hours.
        If Lucy worked for t hours, how many hours did Luke work?

Luke worked for 5 – t hours.
Look at some examples.
If Lucy worked for 2 hours, then Luke would have worked for 3.   5 – 2
If Lucy worked for 1 hour, then Luke would have worked for 4.   5 – 1
If Lucy worked for 4 hours, then Luke would have worked for 1.   5 – 4
If Lucy worked for t hours, then Luke would have worked for 5 – t.

170. If every second seemed like a minute, then how long did the five-minute wait seem like?

If every second seemed like 60 seconds, then five minutes would have seemed like 60 times as long. It would have seemed like 5 × 60 or 300 minutes, which is 5 hours.

171. Originally, her mom had spent $326 on all that stuff. Lucy figured that she could get 17% of that when she sold it. How much is that?

17% of $326
0.17 of 326 = ?
When you know both sides of the *of* you multiply.
0.17 × 326 = 55.42

Lucy sold the dolls, fingernail polish, lipstick, perfume, and doll house and received $55.42.

172. If you are going to get money from others (and are not going to steal it), you have to offer to trade what you have for their money. There are only two things you can offer. Here is the list:

I. Your things.   II. Your labor . This was the first key point of economics: *Your labor is something you sell. It has a price.*

The price of your things depends on how much people want it and what your competition is offering.

The price of your labor depends on how much people want it and what your competition is offering.

Example #1: You offer to sell air. People really want it; they like breathing. But the competition offers it for free—just walk outside and breathe. You can't make money selling air.

Example #2: You offer to sell your old shoe with your signature on it. The competition can't offer that at all. But . . . people don't want your old shoe with your signature on it. You can't make money selling it.

Example #3: You offer your labor of weeding people's gardens. Many people would like that work done, but there are a zillion other people who can do that cheaply. You can't make much money weeding people's gardens.

You want to make money? Offer what people want and at a price that the competition can't beat.

200. It was cold on the first part of one long hike. She drank one cup of water per hour. The second part of the hike was warm, and she drank 3 cups of water per hour. It was warm for 4 more hours than it was cold. She drank a total of 16 cups of water. How long was it cold?

The first step is to notice what was wanted. In this case we want to find out how long it was cold. We could let that equal x or y or z or whatever. When we look for time, I like to use t.

	d cups	r cups per hour	t number of hours
cold			t
warm			

	d cups	r cups per hour	t number of hours
cold			t
warm			$t + 4$

	d cups	r cups per hour	t number of hours
cold		1	t
warm		3	$t + 4$

	d cups	r cups per hour	t number of hours
cold	t	1	t
warm	$3(t + 4)$	3	$t + 4$

It is usually in the last two boxes that are filled in that we can find the equation. We are told that she drank a total of 16 cups.

$$t + 3(t + 4) = 16$$

Distributive law	$t + 3t + 12 = 16$
Combine like terms	$4t + 12 = 16$
Subtract 12 from both sides	$4t = 4$
Divide both sides by 4	$t = 1$

It was cold for 1 hour.

Since we know that it was warm for t + 4 hours and since we know that t = 1, we also know that it was warm for 5 hours. Nobody asked us this, but someday we might be required to give both pieces of information—how long it was cold and how long it was warm.

201. Help Lucy write a list of how to please her railroad customers:

1. My train stations will be neat and clean.
Your handwriting might be better than four-year-old Lucy's.

2. My trains will run on time.

3. My prices will be cheap.

4. All my employees will be very nice to the passengers.

5. Buying tickets will be easy.

6. I won't sell more tickets than there are seats (like some airlines do).

7. There will be lots of free parking at the train stations.

8. My trains will be fast.

9. My trains will be safe.

207. "That will cost you $2,000. Cash on delivery and I'll give you a 14% discount." How much money should Lucy have for that delivery?

If there is a 14% discount, Lucy will be paying 86% of the regular price.

86% of $2,000
0.86 × 2,000
She will be paying $1,720.

208. Are these commutative?
> a) getting some money and
> b) opening a checking account

No. Banks won't let you open a checking account unless you have some money to put into that account. Many banks require a minimum amount to open an account—such as $100. You can't just hand them a dime to start an account.

what dimes used
to look like

217. Let D = the set of all 32 Adventure Duck comic books that Luke bought. Let T = the set of 6 train books that Lucy got from the KITTENS library. Is there a way to create a 1-1 function from D to T?

No. D has too many members. Suppose I tried to make a 1-1 function by assigning:

> the first comic book to the first train book
> the second comic book to the second train book
> the third comic book to the third train book
> the fourth comic book to the fourth train book
> the fifth comic book to the fifth train book
> the sixth comic book to the sixth train book.

The trouble would come when I tried to assign the seventh comic book. I would have to assign it to some train book, but the minute I did that some train book would be the image of two comic books. (Translation: Two elements of the domain would have the same image.) At this point we would no longer have a one-to-one function.

Advanced pre-algebra students would have generalized this problem and have written in their notes: *You can't have a 1-1 function if the domain has more elements than the codomain.*

The super-advanced students would have guessed the theorem*:
If there is a 1-1 function from set A to set B and there is a 1-1 function from set B to set A, then the number of elements in A must equal the number of elements in B. Translation: card A = card B.

The proof of this theorem is not easy. (litotes) I will prove it in *Life of Fred: Five Days of Upper Division Math.* It takes six pages to prove it.　　　　Upper Division = junior/senior in college

218. How many ideas per week did she write down regarding her customers' needs? In the previous problem we had

		ideas per week	
ideas for buying	$20r$	r	20
ideas for customers' needs	$20(r + 5)$	$r + 5$	20

and we found that r was 12.
Therefore, $r + 5 = 17$ customers' needs ideas per week.

———————————————

* Theorems are statements that can be proved. We will prove a lot of theorems in geometry.

227. Almost all of Lucy's 73 dolls each weighed the same. The only exception was Scizzzors, the crocodile doll. He was made out of steel and weighed 864 grams. All 73 dolls together weighed 9,000 grams. How much did each of the 72 non-Scizzzors dolls weigh?

Let x = the weight of one of the non-Scizzzors dolls.
Then 72x = the weight of all the non-Scizzzors dolls.
Then 72x + 864 = the weight of all the dolls.
We know that the weight of all the dolls is 9,000.

$$72x + 864 = 9000$$

Subtract 864 from both sides $72x = 8136$
Divide both sides by 72 $x = 113$

Each of the non-Scizzzors dolls weighed 113 grams.

228. The Lucy Railroad began with $6,000, and suppose it grew at the rate of 2% per month. How much would it be at the end of four months?

Initially, $6,000.
At the end of one month: $1.02 \times 6{,}000 = 6{,}120$
At the end of two months: $1.02 \times 1.02 \times 6{,}000 = 6{,}242.40$
At the end of three months: $1.02 \times 1.02 \times 1.02 \times 6{,}000 \doteq 6{,}367.25$
At the end of four months: $(1.02)^4 \times 6{,}000 \doteq 6{,}494.59$
At the end of four months, Lucy's railroad would have gained almost $500.

229. If the domain is the two parents and the codomain is Lucy and Luke, would this be a function: *Assign each parent to Lucy.*

Yes. Each element of the domain is assigned exactly one element in the codomain. (This function is not one-to-one.)

230. In a pure democracy 51% of the people can make any laws they like.
Make a guess . . . Which of these two documents has the word *democracy* in it?
A) the Declaration of Independence; B) the Constitution; C) both of them; D) neither of them

Neither mention that word. Officially, we are a republic, not a democracy.

232. The cylinder was 10 inches tall and had a radius of 2.4 inches. What is its volume? Use 3.1 for π.

$$V_{cylinder} = \pi r^2 h = (3.1)(2.4)^2(10)$$
$$= 178.56 \text{ cubic inches} \quad (\text{or } 178.56 \text{ in}^3)$$

233. What do you suppose would happen if the government stopped doing health inspections on restaurants?

Some places like **Coalback's Country Cookin'** would try to cut expenses by not cleaning up. They would let the rats and mice run around in the kitchen. They wouldn't wash the dishes very well. His patrons would often get sick—and soon *everyone would know that his place was unclean.*

On the other hand, lots of people would love to know where the clean restaurants are. They would pay to know. Two days after the government got out of the health inspection business, **Sally's Sanitary Spreadsheet** would offer the latest news on which restaurants are clean. Lots of people would *voluntarily* buy her spreadsheet. When the government was in the health inspection business *everyone* was forced to pay taxes to fund the government research.

I, your reader, have a question. How would Sally get a chance to inspect the kitchens of restaurants? She couldn't use government force to demand that restaurants show her those backrooms.

Restaurants—except dirty places like **Coalback's Country Cookin'**—would *love* to have Sally inspect them. They would invite her to inspect them at her earliest convenience. Getting Sally's seal of approval would greatly increase their business.

Government does lots of inspections. If you want to build a house for yourself, in most counties the government demands you get a building permit and have all your plans reviewed by government officials.

In the county I live in, the current waiting time is two and one-half years till you can get permission to build your house on your land.

And people complain about the high price of housing!

If you do any Internet search, there appear to be counties in the United States that don't have zoning or building permits for their non-urban areas. You can build a four-story house made entirely of pizza crusts if you want to.

235. Let's switch things around. Suppose the domain is the set of all numbers and the codomain is the set of all teachers at KITTENS.

Is this a function? *Associate to each number the teacher that is now making that salary each month.*

No. It's not a function for several reasons. First, what would the number 3,989,002,554 be assigned to? No one at KITTENS makes that salary. One part of the definition is that *each element* of the domain must be assigned to something.

Second, two of the teachers might both be 34 years old. Then they would be receive $3,400 as their monthly salary. Another part of the definition of a function is that each element of the domain is assigned to *exactly one* element in the codomain. In this case, 3,400 would be assigned to two different teachers.

Functions is one of the more difficult concepts for many students even though the definition is short: A function is any rule that assigns to each element of the first set exactly one element of the second set.

In Life of Fred: Beginning Algebra we deal with it in Chapter 11.

All of Chapter 7 in Life of Fred: Advanced Algebra is devoted to functions.

In Life of Fred: Trig all of Chapter 2½ looks at functions.

Even when you get to college calculus (Life of Fred: Calculus), all of the first chapter is concerned with one topic: functions.

236. Suppose Aunt Mabel's will left 40% of the waffle iron to you and 12% to each of your other siblings.

How do you handle the sealed bids?

The successful bidder pays their bid to the other five siblings *minus their share* of the waffle iron before the sale.

For example,
Misselthrop got a 12% ownership in the waffle iron. If she had the successful bid of $8, we subtract 12% of that $8 and then divide the remainder according to the will.

You get 40% of $8. . . . . $3.20
Raggledink gets 12% of $8. . . . . $0.96
Widdle gets 12% of $8. . . . . $0.96
Alleshka gets 12% of $8. $0.96
Beattlefoot gets 12% of $8. . . . . $0.96

And that, along with the $0.96 that Misselthrop holds back equals $8.

Aren't you glad your name isn't as weird as your siblings' names?

248. Is there a number that is larger than every number on the number line? We are not talking about ∞, which just means unboundedly large. Is there an actual number?

You won't hear about such a number in beginning algebra, advanced algebra, geometry, trig, or in the first two years of college math (calculus).

Then in upper division (junior/senior) math, we will count the number of natural numbers $\{1, 2, 3, 4, 5, \ldots\}$. We will talk about the cardinality of that set. By that time you will have recovered from the shock of learning about i in advanced algebra.

I don't want to wait. Don't tell me to read faster. I want that number NOW!

Can't you wait until you are older?

I'm 89 years old. I can't wait.

Okay. The cardinal number of $\{1, 2, 3, 4, 5, \ldots\}$ is $\aleph_0$. (Read as aleph-null. *aleph* rhymes with *olivef,* which is olive with an f at the end.)

And the cardinal number of all possible fractions on the number line—which includes $\frac{1}{2}, \frac{11}{100}, \frac{579}{888}, -\frac{7}{8}, \frac{3246841319}{11364888881}$ — is $\aleph_0$.

Hey! I could have guessed that. That $\aleph_0$ is just another number for infinity.

Nope. The cardinal number of the set of all the numbers on the number line is *not equal* to $\aleph_0$. It's a bigger number. It's $\aleph_1$. And in that upper-division math course, we *prove* that $\aleph_0 \neq \aleph_1$. In fact, we prove that $\aleph_0 < \aleph_1$.

This is getting weird. You mean that there are TWO infinite numbers?

Of course not. There are an infinite number of distinct infinite numbers: $\aleph_0, \aleph_1, \aleph_2, \aleph_3, \aleph_4. \ldots$

And we do arithmetic with these transfinite cardinal numbers.

Show me!

$\aleph_3 \times \aleph_4$ equals . . .

Stop! Let me guess. It's $\aleph_{12}$.

Nope.

$\aleph_7$?

Nope. $\aleph_3 \times \aleph_4$ equals $\aleph_4$.

And $\aleph_4 \times \aleph_4 \times \aleph_4 \times \aleph_4 \times \aleph_4 \times \aleph_4 \times \aleph_4 \times \aleph_4 \times \aleph_4 = \aleph_4$.

(continued on next page)

I won't mention that there are many classes of infinite numbers that are larger than any of the alephs: $\aleph_0, \aleph_1, \aleph_2, \aleph_3, \aleph_4$. . . . (Sorry, I couldn't help myself.)

Mathematics is much, much more than just computing 35% of 24.

249. Later that week she took a second hike. She hiked 2 hours in the forenoon and 3 hours in the afternoon. She drank x cups of water per hour in the forenoon and x + 2 cups of water per hour in the afternoon.

She drank a total of 12 cups of water on her hike. In the forenoon how many cups of water per hour did she drink?

	d cups	r cups per hour	t number of hours
forenoon		x	2
afternoon			3

	d cups	r cups per hour	t number of hours
forenoon		x	2
afternoon		x + 2	3

	d cups	r cups per hour	t number of hours
forenoon	2x	x	2
afternoon	3(x + 2)	x + 2	3

We know she drank a total of 12 cups of water.

$$2x + 3(x + 2) = 12$$
$$2x + 3x + 6 = 12$$
$$5x + 6 = 12$$
$$5x = 6$$
$$x = 6/5 = 1.2 \text{ cups of water per hour}$$

250. Lucy could change 5 diapers per hour.
Luke could change 8 diapers per hour.
In 5 hours they changed a total of 34 diapers.
How many hours did Lucy work?

Let t = the number of hours that Lucy worked.
Then 5 – t = the number of hours that Luke worked. (We did this in the previous problem.)
Then 5t = the number of diapers that Lucy changed. (She had changed 5 diapers per hour for t hours.)
Then 8(5 – t) = the number of diapers that Luke changed. (He had changed 8 diapers per hour for 5 – t hours.
Then 5t + 8(5 – t) = the total number of diapers that they changed.

$$5t + 8(5 - t) = 34$$

Distributive law	$5t + 40 - 8t = 34$
Combine like terms	$-3t + 40 = 34$
Add 3t to both sides	$40 = 34 + 3t$
Subtract 34 from both sides	$6 = 3t$
Divide both sides by 3	$2 = t$

Lucy worked for 2 hours.

You will learn that 5t – 8t equals –3t in algebra. A gain of 5 and a loss of 8 is a loss of 3.

251. Lucy found six books on trains at the KITTENS library and stuffed them into her 0.58-pound backpack. The books all weighed the same. She handed the 10 pounds to her brother to carry. How much did each book weigh?

Let x = the weight of one book.
Then 6x = the weight of all six books.
Then 6x + 0.58 = the total weight.

$$6x + 0.58 = 10$$

Subtract 0.58 from both sides	$6x = 9.42$
Divide both sides by 6	$x = 1.57$

Each book weighed 1.57 pounds.

Writing 0.58 instead of just .58 can often help avoid mistakes. With just .58, it is too easy to miss the decimal point.

252. The wealth of a community is owned by the individuals (or families) who have created that wealth. If Lucy's father works at a tire store, he trades his labor for a paycheck.

Governments often treat all the money in the community as if it were owned by the community collectively. It takes (taxes) money from the individuals and uses it for the "common good."

It would be fair if: ① each person was taxed the same as every other person, and ② everyone received the same benefit from the government spending.

Which of these two (① or ②) is not true?

The wealth of the individuals in the community is treated as the Commons. The government puts its "cows" on the common field (taxes).

① Does it tax everyone equally? No. The original Constitution (1789) of the United States stated that everyone was to be taxed equally.* (That was changed 124 years later in the 16th Amendment, which allowed unequal taxation.)

② Does everyone receive the same benefit? Special interest groups (especially ones with good lobbyists) receive most of the money.

Just do an Internet search "government wasteful spending." My favorite is the $75.5 million ($75,500,000) that was spent to build an airport in the town of Akutan, Alaska. The town has 75 full-time residents. The airport has no roads to it. The airport has no airlines serving it. The town itself has no electricity or running water.

Ha! Ha! Ha!

253. If it took her $5\frac{1}{3}$ days to do the whole job, what fraction would be done each day?

Look at what you did in the previous problem.
The whole job in 5 days → 1/5 of the job each day.
The whole job in 4 days → 1/4 of the job each day.
The whole job in 10 days → 1/10 of the job each day.
So the whole job in 5⅓ days → 1/5⅓ = 1 ÷ 5⅓ = 1 ÷ 16/3 = 1 × 3/16 = 3/16 of the job done each day.

---

* Article 1, Section 9, Paragraph 4: "No . . . direct tax shall be laid, unless in proportion to the census or enumeration. . . ."

257. If you invert $\frac{2}{3}$ and $\frac{3}{4}$ which one is smaller?

Which is smaller? $\frac{3}{2}$ or $\frac{4}{3}$

Again, to compare fractions, you *pretend* that you are going to add them. You make their denominators alike.

$$\frac{3}{2} = \frac{9}{6} \qquad\qquad \frac{4}{3} = \frac{8}{6} \qquad\qquad \frac{4}{3} < \frac{3}{2}$$

260. When Lucy got her first train station, she was really happy. She might have assigned 1400 **HaPPiNeSS PoiNtS** to that event. When she gets her 485th train station, do you imagine that she will assign 1400 **HaPPiNeSS PoiNtS** to that?

Probably not. After you have had one spoonful of your favorite ice cream, the thirtieth spoonful doesn't have the same impact. After eating a gallon of your favorite flavor, there might be **pain** involved in eating another spoonful.

The **HaPPiNeSS PoiNtS** you assign are not constant. The games you enjoyed playing five years ago may be different than the ones you like now.

When you get married, the traditional vows often include the promise that you will stick with that person "in sickness and in health, in riches or in poverty." What that means is that you will remain married regardless of <u>your current</u> assignment of **HaPPiNeSS PoiNtS** to being with that person.

"Being in love" (10,000 **HaPPiNeSS PoiNtS**) is a lousy reason to get married. That usually lasts only a year or two.

Loving someone, in contrast, can last a lifetime.

261. Plot $y = 10x^2 + 7$ from x = 0 to x = 6.

If x = 0, then $y = 10x^2 + 7$ means y = 7.   (0, 7)

If x = 1, then $y = 10x^2 + 7$ means y = 17.   (0, 17)

If x = 2, then $y = 10x^2 + 7$ means y = 47.   (0, 47)

If x = 3, then $y = 10x^2 + 7$ means y = 97.   (0, 97)

If x = 4, then $y = 10x^2 + 7$ means y = 167   (0, 167)

If x = 5, then $y = 10x^2 + 7$ means y = 257   (0, 257)

If x = 6, then $y = 10x^2 + 7$ means y = 367   (0, 367)

273. When people are going to trade their money for the Two Things You Can Offer (see the previous question), they are looking at two things: ① the price and ② the quality. (This is called the *second key point of economics*.)

Lucy thought of being a carpenter. Looking at the second key point of economics, why wouldn't that work?

The second key point of economics = price and quality. Even if Lucy offered to work for 1¢ per hour, the quality of her work wouldn't be good enough. Would you pay Lucy to build your house? She couldn't even drive nails using a 32-ounce hammer.

No one would hire Lucy as a secretary, even if she would work for 1¢ per hour. She doesn't know how to type or even read.

Some jobs require physical skills. Some require education. Lucy had neither. No one would trade their money for her labor.

When you are a kid, you have two main things to do: ❶ Grow up (physical skills) and ❷ Learn stuff (education). When you have done these things, the quality of your labor has gone up. You have something to offer the world.

Right now, just stay cute (and obedient). You can offer that to your parents, and they will offer you free food and a place to stay at their house. They might even throw in some kisses and hugs for free.

274. Given these two facts—① you need lots of people with different skill sets and ② you shouldn't share ownership—what should you do?

Easy. You hire them.

Small discussion:
You hire them. You guide them and let them know what is required of them. You love them. You may need to fire some of them. Even if you are good at hiring, the Pareto principle says that some of them may be duds.

PARTNERS CANNOT BE FIRED.

275. The government could get 100,000 toothpicks.

It could dye some of the red. Those would be 1-picks.
Some of them blue. Those could be 10-picks.
Some, green. Those could be 100-picks.
Some orange. 1,000-picks.
Some purple. 10,000-picks.

Why would this be a bad money system?

Certainly, they would be durable and portable.

They would be fungible. Any two green toothpicks would have the same value.

They would be divisible. Sixty-seven orange would be worth 67,000 red.

The only real question is that of limited supply. **Hey! I, your reader, have read your question. You said that the government would only get 100,000 toothpicks. That's not an infinite number.**

That's true. But in the real world there are two things that could mess up the property of limited supply.

❶ I could go buy a bunch of toothpicks and some orange dye and create an unlimited number of 1,000-picks. This is called **counterfeiting**.*

❷ The government could increase the money supply by creating a million new pieces of toothpick money. This is called **inflation**.

small essay
Inflation

Inflation is when the government increases the supply of money that it creates. When there are ten times as many colored toothpicks in circulation, then they tend to become worth less than they did before. Prices will rise because of inflation.

Inflation is not defined as rising prices. It is defined as increasing money supply.

Years ago President Gerald Ford appeared before Congress and declared that inflation was "public enemy number one." Since the government is the only one that can increase the money supply and create inflation, that would mean that the government itself is public enemy number one.

Even Alan Greenspan labeled Whip Inflation Now as: "This is unbelievably stupid" in his book *The Age of Turbulence*.

---

* I before E, except after C—except in English that has a zillion exceptions.

On the day of Ford's speech to Congress, the Whip Inflation Now form was made available. It read, "Dear President Ford: I enlist as an inflation fighter. . . . I will do the very best I can for America."

If you signed that form and sent it in, you could get your very own plastic button.

this button really existed

Some people really believed that nonsense.

It is government alone that legally controls the number of toothpicks, the number of pieces of green paper they call dollars, that are in circulation.

When I was in college in the 1960s, one of my upper-division math textbooks cost $5.00 (new). Another cost $4.95 (new). Now they are each priced at well over $100.

When the government prints up too much paper money, it can look like this photo from Hungary in 1946.

Or more recently, in 2008 you could own a

worthless paper money

one hundred trillion dollar bill issued by Zimbabwe. It was the government of Zimbabwe that created that inflation by increasing their money supply.

You want to stop inflation? That's easy. Before President Nixon, the government was required by law to exchange the paper money they printed for silver or gold. If they printed too much, people would turn in the paper and the government would run out of silver or gold.

Today, it's just paper.

end of small essay

276. Each month it was worth 8% more than the previous month. Approximately how long would it take for that worth to double?

By the Rule of 72, it would take $\frac{72}{8}$ = 9 months.

300. Ginny's head was 5 cm (cm = centimeter) tall. That was two-ninths of his total height. How tall was Ginny?

The conversion factor will be $\dfrac{5 \text{ cm}}{2/9 \text{ total height}}$

We want to convert Ginny's total height into centimeters.

$$\frac{\text{Ginny's total height}}{1} \times \frac{5 \text{ cm}}{2/9 \text{ total height}}$$

$= \dfrac{5 \text{ cm}}{2/9}$

$\boxed{5 \div 2/9 \;=\; 5 \times \dfrac{9}{2} \;=\; \dfrac{45}{2} \;=\; 22\tfrac{1}{2}}$

$= 22\tfrac{1}{2}$ cm

For those who don't think in the metric system:
5 cm is almost exactly 2 inches.
More precisely, 5 cm ≈ 1.9685 inches.

≈ means "approximately equal to."

301. If the domain is {toot} and the codomain is {thank you, good morning, please pass the salt}, is this a function? *Assign to toot to thank you; assign toot to good morning; assign toot to please pass the salt.*

The definition of a function states that each member of the domain must be assigned to <u>exactly one</u> member of the codomain. Since toot is assigned to three different members of the codomain this is not a function .

302.
$$\begin{array}{r} 839 \\ + \ 764 \\ \hline 1603 \end{array} \qquad \begin{array}{r} 3247 \\ - \ 1889 \\ \hline 1358 \end{array}$$

303. If Lucy were to offer a job to do that physical work for $18/hour, which of these people might be willing to accept?

A) Lucy's dad who is currently making $26/hour as an assistant manager at **Tammy's Tires**

B) Lucy's mom who is currently making $13/hour working in a candy store

C) Lucy's brother Luke who makes $5/hour whenever he does extra chores around the house.

B or C is the answer. Each of those have opportunity costs that are less than the $18/hour that Lucy is offering.

304. Imagine working 50 feet off the ground on high-voltage electrical lines. That would scare me silly. What would happen if everyone assigned 0.0002 ***HaPPiNeSS PoiNtS*** to that work?

  If that work didn't get done, then in a matter of weeks houses would start to go dark. No television. No dishwasher. No lights.

  I am grateful that there are individuals who enjoy that kind of work. Many of them would hate to work in an office stapling pieces of paper together and sticking things in file cabinets.

  Imagine if no one enjoyed taking care of babies.

  Imagine if no one got ***HaPPiNeSS PoiNtS*** making pizzas. Horrors! That last thought is too horrible to contemplate.

  We each assign different ***HaPPiNeSS PoiNtS*** to different activities. If we didn't, the world would be a mess.

305. Plot (1, 0), (2, 2), (3, 5), and (4, 9).

Lucy's panic level was skyrocketing.

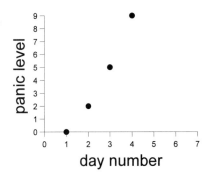

306. Make the production chart.

	🍽️	🛏️
Lucy's mom	200	12
Luke	40	4

308. Lucy looked at the cans of spaghetti at Alfredo's Foods. A large can of Alfredo brand spaghetti cost 48¢ per can.

Alfredo brand spaghetti also comes in a smaller can. Sixteen of those smaller cans have the same amount as 12 of the bigger cans.

What would be the expected price of a smaller can?

There are several ways to do this problem.

One way:

Twelve large cans @ 48¢ each would cost 576¢.

Sixteen smaller cans should each cost $\dfrac{576¢}{16 \text{ small cans}}$ which is 36¢/small can.

If the small cans are less than 36¢ each, Lucy should buy the small cans. If the small cans are more than 36¢, Lucy should buy the large cans.

A second way:

Six pretty boxes.

	total cost	cost/can	number of cans
large cans	48(12)	48	12
small cans	16x	x	16

$$16x = 48(12)$$
$$16x = 576$$
$$x = 36¢ \text{ per small can}$$

A third way:

Use a conversion factor.

$$\frac{48¢}{\text{large can}} \times \frac{12 \text{ large cans}}{16 \text{ small cans}} = \frac{36¢}{\text{small can}}$$

309.

$$\begin{array}{r} 78 \\ \times\ 64 \\ \hline 312 \\ 468 \\ \hline 4992 \end{array} \qquad \begin{array}{r} 869 \\ \times\ 70 \\ \hline 000 \\ 6083 \\ \hline 60830 \end{array}$$

Complete Solutions and Answers

311. Lucy's train from Reno to Sacramento will travel at 30 mph. When it goes back from Sacramento to Reno it will go 20 mph since it will be traveling uphill and will take 2 hours longer. How long will it take to go from Reno to Sacramento?

       Let t = the time from Reno to Sacramento.

       Then t + 2 = the time it takes to go from Sacramento to Reno.

       Then 30t = the distance from Reno to Sacramento.*

       Then 20(t + 2) = the distance from Sacramento to Reno.

Since these two distances are the same, 30t = 20(t + 2).

312. Do

       Lucy could change 5 diapers per hour.
       Luke could change 8 diapers per hour.
       In 5 hours they changed a total of 34 diapers.
       How many hours did Lucy work?

using six pretty boxes.

	diapers changed	rate	time
Lucy			t
Luke			

	diapers changed	rate	time
Lucy			t
Luke			$5 - t$

	diapers changed	rate	time
Lucy		5	t
Luke		8	$5 - t$

---

* If it takes t hours and you are going at 30 mph, then the distance is equal to rate times time (d = rt), which is 30t.

	diapers changed	rate	time
Lucy	$5t$	5	t
Luke	$8(5-t)$	8	$5-t$

Since the total number of diapers changed was 34,
we have $5t + 8(5 - t) = 34$. We did the solution of this equation in the previous problem.

313. 1½% times 12 $= \dfrac{3}{2}\% \times 12 = \dfrac{36}{2}\% = 18\%$ per year

314. Suppose that the first set (the domain) is the set of students at KITTENS University. Suppose the second set is the set all humans. Why is this not a function: *Assign to each student their favorite ice hockey player.*

  Not every student has a favorite ice hockey player. Can you imagine that? Not *every element* of the domain is assigned something in the codomain.

315. For many students is the concept of function one of the hardest topics they have met in mathematics? ⊠ yes ⊠ yes

316. The domain: Ginny, the engineer, 2 pig dolls—Porky and Ima, 68 girl dolls, 1 boy doll, and 1 alligator doll named Scizzzors.
  The codomain: 1 red engine, 1 green train car, 1 blue train car, 1 brown train car, and 50 gray train cars.
  Is this a function? *Lucy put Ginny in the engine, the pig dolls in the brown car, the 68 girl dolls in the green car, and the boy doll in the blue car. (She left the 50 gray cars empty.)*

  There are two parts to the definition of a function. First, did every member of the domain get assigned? Yes. **Stop! I, your reader, object. You forgot Scizzzors.** You're right. I goofed. This is not a function.

319. By noon 1,000 people had dined at Lucky Lucy's Lounge. Some of them had paid in cash. Some of them had ordered ribs. Some of them had done both; they ordered ribs and paid in cash. Draw a Venn diagram.

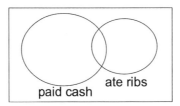

320. Try out different values of x (from the previous problem) and find out when Lucy will be a billionaire. Translation: When she will be worth more than $1,000,000,000.

If x = 6, then y = 10x^2 + 7 means y = 367
If x = 7, then y = 10x^2 + 7 means y = 497
If x = 8, then y = 10x^2 + 7 means y = 647
If x = 9, then y = 10x^2 + 7 means y = 817
If x = 10, then y = 10x^2 + 7 means y = 1007

We were told that y is measured in millions of dollars, so y = 1007 means 1007 millions of dollars. That is $1,007,000,000.

By then she will have expanded her empire to include all 50 states.

For fun, here's the graph:

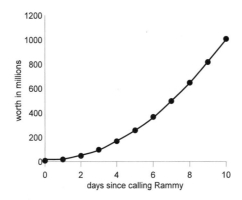

327. Lucy's goal in life was to create the Lucy Railroad. The definition of *tool* is something that helps you accomplish your goal more easily. In a sentence or two explain why learning lots about trains is a tool for Lucy.

One sentence: The more that Lucy learned about trains, the more easily she could do the things needed to create her railroad when she grew up.

Education is a tool that is super valuable in today's world. Two hundred years ago having a good back and strong hands were important tools.

Today, how often do you see "Help Wanted" ads for bodies. They want minds (=·experience and education).

328. Lucy has 36 dolls that she would be happy to get rid of.
She could trade 3 dolls for 2 hammers.
She could trade 4 hammers for 5 screwdrivers.
She could trade 6 screwdrivers for 5 train cars.
Convert those 36 dolls into train cars.

$$\frac{36 \text{ dolls}}{1} \times \frac{2 \text{ hammers}}{3 \text{ dolls}} \times \frac{5 \text{ screwdrivers}}{4 \text{ hammers}} \times \frac{5 \text{ train cars}}{6 \text{ screwdrivers}}$$

$$= \frac{36 \text{ dolls}}{1} \times \frac{2 \text{ hammers}}{3 \text{ dolls}} \times \frac{5 \text{ screwdrivers}}{4 \text{ hammers}} \times \frac{5 \text{ train cars}}{6 \text{ screwdrivers}}$$

$$= 25 \text{ train cars}$$

329. Solve $\qquad\qquad 9y - 13 = 3y + 17$

Add 13 to both sides	$9y = 3y + 30$
Subtract 3y from both sides	$6y = 30$
Divide both sides by 6	$y = 5$

330. Each little bite of cereal is a little locomotive that is 95% oats and 5% engine oil. A handful of Toot!® (130 grams) would contain how much engine oil?

Five percent of 130 grams
5% of 130 = ?
When you know both sides of the *of*, you multiply.
$0.05 \times 130 = 6.5$

Each handful of Toot!® contains 6.5 grams of engine oil.

331. Which one of these would be hardest for Lucy to learn using only books and the Internet? A) U.S. history; B) great literature; C) chemistry; D) writing good English; E) philosophy

My guess is: D) writing good English. Certainly, the ability to communicate well in English can be helped by reading books that teach you how to punctuate sentences correctly and how to avoid stylistic errors.* The thing that books can't offer is the practice in writing English that is read and criticized by others.

I, your reader, think that C) chemistry was also a good answer. A real part of chemistry is fooling around in a lab. While you can learn to balance chemical equations from a book, how are you going to learn how to pour liquids from one test tube to another by looking in a book?

Learning how to communicate clearly is *absolutely critical* in almost every human occupation. The only exception that I can think of is being a hermit. On the other hand, washing test tubes or reading dials on a chemical analyzer are tasks that super easy to pick up when you need them, in contrast to learning the complexities of working with 4,5-disecbutyl-2,46-nontriyne. (← see page 223 of *Life of Fred: Chemistry*.)

332. If you are 3, what percent increase is needed to become 16?

To go from 3 to 16 is an increase of 13.
13 is what percent of 3?   13 = ?% of 3
$13 \div 3 \approx 4.3333333 = 433.33333\% \doteq 433\%$

To get to be 16 will require that Lucy become 433% older than she is now. *When you are 3, becoming 16 seems like forever.*

---

* For example, prefer complex sentences over compound sentences. Compound sentence = "John bought the car, and Mary went broke." Complex sentence = "Because John bought the car, Mary went broke."

In good writing about 6% of the sentences are compound while about 44% are complex. Compound sentences can become boring.

Complex sentences are more varied and more exact. Complex sentences often use these subordinating conjunctions: after, although, as, as long as, because, before, if, so that, though, unless, when, where, while.

333. One pound is approximately equal to 454 grams. In the previous problem we found that one of the non-Scizzzors dolls weighed 113 grams. Does a quarter-pound hamburger weigh more or less than one of these dolls?

There are two ways we might do this problem.

Convert everything into pounds

$$\frac{113 \text{ g}}{1} \times \frac{1 \text{ lb.}}{454 \text{ g}} \approx 0.248898 \text{ lbs.}$$

($\approx$ means "approximately equal to)

The doll weighs 0.248898 lbs. The hamburger weighs 0.25 lbs. The hamburger weighs more than the doll.)

> How did we get the 0.25?
> In *Life of Fred: Decimals and Percents* you memorized the nine conversion factors. One of them was 1/4 = 25%, which is 0.25.
>
> If you didn't memorize that, you would have to divide 1 by 4.
> $$4\overline{)1.00}^{\,0.25}$$

Convert everything into grams

$$\frac{0.25 \text{ lbs.}}{1} \times \frac{454 \text{ g}}{1 \text{ lb.}} = 113.5 \text{ grams}$$

The doll weighs 113 grams. The hamburger weighs 113.5 grams. The hamburger weighs more than the doll.

Those two things are almost the same weight. 113 grams ≈ ¼ lbs. When you are at the restaurant, instead of ordering a quarter-pound hamburger, you could order a 113-gramburger. (I just made up that word.)

334. We want education to be *fast*, cheap, and pleasant.
Let's start with pleasant. Which is these is most pleasant?
A) Reading
B) Watching movies, television, and videos
C) Going to lectures and talks

My best guess is B) Watching movies, television, and videos. You can sit at home and eat popcorn and spend hours watching colorful and dramatic scenes of railroad adventures. It's fun.

335. Lucy wrote out 43 items on her **Action List**. She also spent 19 seconds scratching some of the pink paint out of her hair. It itched. Altogether, she spent 750 seconds on these two things. How long did it take her to write each item?

Let x = the amount of time it took Lucy to write one item.
Then 43x = the amount of time it took to write all 43 items.
Then 43x + 19 = the total amount of time spent.

$$43x + 19 = 750$$
$$43x = 731$$
$$x = 17$$

It took 17 seconds to write each item.

336. Any of these would slow her down.

337.
```
        276
  84)23184
     168
     638
     588
     504
     504
```

338. Yes.*

339. If there were 18 pennies in each cubic inch, how many pennies were in Lucy's cylinder?

Using a conversion factor . . .

$$\frac{178.56 \text{ cubic inches}}{1} \times \frac{18 \text{ pennies}}{1 \text{ cubic inch}}$$

$$= 3,214.08 \doteq 3,214 \text{ pennies}$$

---

* There are other possibilities: Ouch! No. Who? Perhaps. Why?

340. The station needed a little electrical work. She worked for 6 hours in the forenoon and installed 7 electrical outlets per hour. In the afternoon she worked for 8 hours. That day she installed a total of 90 outlets. At what rate was she installing outlets in the afternoon?

	d outlets installed	r outlets per hour	t hours
forenoon			
afternoon		x	

	d outlets installed	r outlets per hour	t hours
forenoon		7	6
afternoon		x	8

	d outlets installed	r outlets per hour	t hours
forenoon	42	7	6
afternoon	$8x$	x	8

Since she installed a total of 90 outlets . . .

$$42 + 8x = 90$$
$$8x = 48$$
$$x = 6$$

In the afternoon she installed 6 outlets per hour.

341. The two dresses and the $8.08 jewelry all cost $45.

Let x = the cost of one dress.
Then $2x$ = the cost of both dresses.
Then $2x + 8.08$ = the total cost.

$$2x + 8.08 = 45$$
$$2x = 36.92$$
$$x = \$18.46 \leftarrow \text{cost of one dress}$$

350. Let set H = {1, 2, 3, 4, . . . , 23, 24}. Let set A = {sleeping, reading, talking with people connected with railroading, planning}.

      She assigned each element of H to an element of A.

1 → sleeping; 2 → sleeping; 3 → sleeping; . . . ; 10 → reading; etc.

      This is a function. Is it one-to-one?

      No. Both 1 and 2 are assigned to sleeping.

351. Plot $y = x^3$ for values of x between 0 and 3.

      To graph any equation you plot some points and then you draw the curve through those points.

      To plot a point you first name a value for x. Then use the equation to find the corresponding value of y.

      If x = 0, then $y = x^3$ becomes $y = 0^3$, which is 0.   (0, 0)
      If x = 1, then $y = x^3$ becomes $y = 1^3$, which is 1.   (1, 1)
      If x = 2, then $y = x^3$ becomes $y = 2^3$, which is 8.   (2, 8)
      If x = 3, then $y = x^3$ becomes $y = 3^3$, which is 27.   (3, 27)

If I wanted more points, I might let x = 2.5. Then $y = x^3$
becomes $y = 2.5^3$ which is 15.625.   (2.3, 15.625)

Then plot the points.

Then draw the curve.

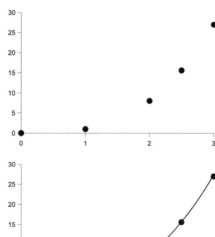

355. Take two positive numbers, x and y, where x < y. Is it always true that $x^2 < y^2$?

One way to approach this problem is to try lots of different examples.

2 < 5 and 4 < 25 is also true.

10 < 40 and 100 < 400 is also true.

$\frac{1}{4} < \frac{2}{3}$ and $\frac{1}{16} < \frac{4}{9}$ is also true.

0.2 < 0.6 and 0.04 < 0.36 is also true.

It seems to be always true. In symbols: $0 < x < y \Rightarrow x^2 < y^2$.

"$\Rightarrow$" is the logic symbol for "implies."

In algebra we can prove that $0 < x < y \Rightarrow x^2 < y^2$, but we don't have the background yet to do it here.

356. In an hour Lucy's mom can either wash 200 dishes or make 12 beds. For each bed that she makes, how many dishes does she give up washing? (Translation: What is the opportunity cost for making a bed?)

To make 12 beds she avoids washing 200 dishes.

To make a bed she avoids washing $\frac{200}{12}$ dishes.

357. When Lucy dies, she wants to leave her business to her six kids. Businesses are a little like waffle irons—they can't be cut up into six parts. Sealed bids by the kids are a lot better than fistfights. Lucy will put a "sealed bids" clause in her will.

But there is one drawback. With Aunt Mabel's waffle iron, all the nephews and nieces each had some spare money to bid with.

None of Lucy's kids have the millions it would take to make a bid. What else might Lucy put in her will so that her kids can make bids?

Here is one thought . . .

She should include a clause in her will that specifies that the bids will be in the form of a promissory note that reads: "I promise to pay to my five other siblings the amount of [insert five-sixths of the bid amount here] payable in monthly payments of [insert 1% of the bid amount here] or more until it is all paid."

The winning bidder will be able to pay on this note using the income from the business.

390. Lucy learns 7 new railroad ideas in every 12 minutes from the reading that she does. In 88 minutes how many ideas might she expect? (Round your answer to the nearest whole number.)

Using a conversion factor,

$$\frac{88 \text{ minutes}}{1} \times \frac{7 \text{ ideas}}{12 \text{ minutes}} = 51\tfrac{1}{3} \doteq 51 \text{ ideas}$$

We needed to round the number of ideas because the number of ideas is a discrete quantity. You can't have a third of an idea.

391. Luke took Lucy's 10-pound backpack and put his 32 comic books in it. The whole thing now weighed 22 pounds. The comic books all weighed the same. How much did each comic book weigh? Do this problem using fractions instead of decimals.

Let x = the weight of a comic book.
Then 32x = the weight of all the comic books.
Then 32x + 10 = the total weight that Luke was carrying.

$$32x + 10 = 22$$

Subtract 10 from both sides $\qquad 32x = 12$

Divide both sides by 32 $\qquad x = \dfrac{12}{32}$

$\dfrac{12}{32} = \dfrac{6}{16} = \dfrac{3}{8}$ Each comic book weighed three-eights of a pound.

392. If x is an even number, what is the next consecutive even number?

If x were, for example, 16, then the next consecutive number would be 18.

If x were, for example, 104, then the next consecutive number would be 106.

The next consecutive even number after x is x + 2.

393. On one of the days that Lucy was working on her train station she put in t hours doing electrical work, $t - 4$ hours doing plumbing, and $t - 2$ hours doing painting. Electrical work increased the value of her train station by \$16/hour. Plumbing, \$18/hour. Painting, \$12/hour.

On that day she increased the value of the station by \$180. How long did she do electrical work?

	increase in value	rate	time
electrical			t
plumbing			$t - 4$
painting			$t - 2$

	increase in value	rate	time
electrical		16	t
plumbing		18	$t - 4$
painting		12	$t - 2$

	increase in value	rate	time
electrical	$16t$	16	t
plumbing	$18(t - 4)$	18	$t - 4$
painting	$12(t - 2)$	12	$t - 2$

The total increase in value was \$180: $16t + 18(t - 4) + 12(t - 2) = 180$

Distributive law	$16t + 18t - 72 + 12t - 24 = 180$
Combine like terms	$46t - 96 = 180$
Add 96 to both sides	$46t = 276$
Divide both sides by 46	$t = 6$

Lucy worked 6 hours doing electrical work.

400.   That afternoon she called **KAMMY'S KWICK KITCHENS** and asked them to install a kitchen tomorrow morning at the I Love You Amy Station.

She called her receptionist and asked her to hire some cooks and to print up some menus.

She asked her mom to hire some people to serve the food.

She asked her brother Luke to arrange for some music.

If the domain = {Kammy's Kwick Kitchens, receptionist, mom, Luke} and the codomain is all the things that needed to be done, then this is *not* a function. Why not?

The definition of a function is any rule that assigns to each element of the domain **exactly one** element in the codomain. The receptionist was assigned to two different tasks.

401. Convert $\frac{3}{8}$ pound into ounces. (1 pound = 16 ounces)

$$\frac{3 \text{ pound}}{8} \times \frac{16 \text{ ounces}}{1 \text{ pound}}$$

$$= \frac{3 \ \cancel{\text{pound}}}{8} \times \frac{16 \text{ ounces}}{1 \ \cancel{\text{pound}}}$$

$$= \frac{3}{\cancel{8}_1} \times \frac{\cancel{16}^2}{1} = 6 \text{ ounces}$$

402.

$$\frac{3}{5} + \frac{2}{3} \qquad\qquad \frac{7}{8} - \frac{1}{3}$$

$$\frac{9}{15} + \frac{10}{15} \qquad\qquad \frac{21}{24} - \frac{8}{24}$$

$$\frac{19}{15} \qquad\qquad\qquad \frac{13}{24}$$

$$1\frac{4}{15}$$

403. Here is how we might use that Venn diagram that you just drew. Suppose 700 had paid in cash and 400 had ordered ribs. Suppose 100 had done both.

How many had paid in cash but didn't order ribs?

The only number—of the 700, 400, and 100—that we can start with is the 100.

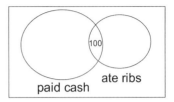

After putting in the 100, then we know that 600 (= 700 − 100) is the next number we can put on the diagram.

Six hundred paid cash but didn't order ribs.

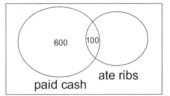

404. Again, let the domain be the set of students at KITTENS and the codomain be the set of all humans. Is this a function? *Associate to each student their biological mother.*

Yes. This is a function.

One moment! I, your reader, object. This couldn't be a function.
Why not?
Fred is a human. Fred is in the codomain.
I agree.
No student has Fred as a mother.
That's also true. We have to read the definition of function carefully. It says that a function is any rule that assigns to each element of the domain exactly one element of the codomain.

So each element of the domain will be involved. It doesn't talk about every element of the codomain.

Every KITTENS student will have one mom. It's a function.

To get technical for a moment, we note that the set of moms of KITTENS students is a subset of the codomain, which is the set of all humans.

145

440. Is the number of Fred Gauss dolls a discrete variable or a continuous variable?

It is a discrete variable. It could be 0 or 1 or 2 or 89, but it couldn't be $1\frac{7}{8}$ or 3.4.

A variable is continuous if it can take on the values of, say, 7 and 8, then it can take on *every* value between 7 and 8.

441. Make the opportunity cost chart.

Production Chart

Lucy's mom	200	12
Luke	40	4

Opportunity Cost Chart

Lucy's mom	$\frac{12}{200}$	$\frac{200}{12}$
Luke	$\frac{4}{40}$	$\frac{40}{4}$

452. Lucy has a collection of 2,307 cards. On each card she has written down one idea that she has gotten from her reading.

Let those cards be the domain.

She has 18 boxes, each with a label on it. The codomain is those 18 boxes. Lucy puts each card in a box. This is a function.

Your question: Is this function one-to-one?

If a function is 1-1, then no two elements of the domain can be assigned to the same element in the codomain.

With 2,307 cards stuffed into 18 boxes, at least one of those boxes must contain at least two cards. This function is not one-to-one.

453. What is the interest charge of $1\frac{1}{2}\%$ on $12,000?

$1\frac{1}{2}\%$ of 12,000 = ?
You know both sides of the *of,* so you multiply.
$1\frac{1}{2}\% \times 12,000$
$1.5\% \times 12,000$
$0.015 \times 12,000$
$180

How many days of work would it take you each month to pay that $180 of interest? Buying nonproductive things with *money you have saved*, rather than money you have borrowed, can add days to your life.

454. $4\frac{1}{5} \times 7\frac{1}{8} = \frac{21}{5} \times \frac{57}{8} = \frac{1197}{40} = 29\frac{37}{40}$

$$
\begin{array}{r}
29 \text{ R } 37 \\
40)\overline{1197} \\
\underline{80} \\
397 \\
\underline{360} \\
37
\end{array}
$$

147

455. The bottle was $\frac{9}{10}$ full when Lucy started painting the car. When Lucy was done, there was $\frac{1}{6}$ of the bottle left. How much had she used in painting the train car pink?

Using the General Rule,* if the bottle started with 10 ounces and there was 2 ounces left after Lucy was done, she would have used 8 ounces. We subtracted.

$$\frac{9}{10} - \frac{1}{6}$$

We could use a common denominator of 60, but that would involve more arithmetic than if we used the least common denominator of 30.

$$\frac{27}{30} - \frac{5}{30} = \frac{22}{30} = \frac{11}{15}$$

Lucy had used eleven-fifteenths of the bottle of nail polish.

456. Solve
$$12x - 4x = 5$$
Combine like terms
$$8x = 5$$
Divide both sides by 8
$$x = \frac{5}{8}$$

457. Scizzzors weighs 864 grams. A regular doll weighs 113 grams. To the nearest percent, how much more does Scizzzors weigh than a regular doll?

Scizzzors weighs 751 grams more than a regular doll. (864 – 113)
751 is what percent more than 113?
751 = ?% of 113
We don't know both sides of the *of* so we divide the number closest to the *of* into the other number.

$$751 \div 113 \approx 6.646 = 664.6\% \doteq 665\%$$

---

* The General Rule says that if you don't know whether to add, subtract, multiply, or divide, use some simple numbers first and see which operation you used.

460. Solve $\qquad$ $30t = 20(t+2)$

Distributive law $\qquad$ $30t = 20t + 40$

Subtract 20t from both sides $\qquad$ $10t = 40$

Divide both sides by 10 $\qquad$ $t = 4$

It would take 4 hours to go from Reno to Sacramento.

461. Seven machines plus $5.09 for shipping will cost $55. How much is each machine?

Let x = the cost of a machine.
Then 7x = the cost of seven machines.
Then 7x + 5.09 = the total cost.

$\qquad$ $7x + 5.09 = 55$

Subtract 5.09 from both sides $\qquad$ $7x = 49.91$

Divide both sides by 7 $\qquad$ $x = 7.13$

Each machine will cost $7.13.

462. Let C = {purple, yellow}. Let J = the set of the 31 jockey hats that Fred has.

Is it possible to create a function from C to J that is *not* 1-1?

Yes. For example, *assign purple and yellow to Fred's 5th day hat with the feather on it.*

The 2% of pre-algebra students that are in the super advanced category can tell me how many possible functions there are from C to J that are not 1-1. I'll put the answer to this in #912.

463. The normal installation charge was $1,036, but Lucy needed everything in a hurry. She chose to pay the extra 24% fee for fast service. How much was the total bill?

The long way to do "24% more" problems:
24% of $1,036 is $248.64.
$248.64 plus the original $1,036 is $1,284.64.
The easier way:
24% plus the original 100% is 124%
124% of $1,036 is $1.24 \times 1,036 = $1,284.64.

466. $\dfrac{3}{4} \div 2\dfrac{2}{5}$

$\dfrac{3}{4} \div \dfrac{12}{5}$

$\dfrac{3}{4} \times \dfrac{5}{12}$

$\dfrac{{}^1\cancel{3}}{4} \times \dfrac{5}{\cancel{12}_4}$

$\dfrac{5}{16}$

$8\dfrac{1}{8} - 4\dfrac{1}{4}$

$8\dfrac{1}{8} - 4\dfrac{2}{8}$

$7\dfrac{9}{8} - 4\dfrac{2}{8}$

$3\dfrac{7}{8}$

467. $(5\dfrac{1}{6})^3 = \dfrac{31}{6} \times \dfrac{31}{6} \times \dfrac{31}{6} = \dfrac{29791}{216} = 137\dfrac{199}{216}$

My calculator helped a lot on this problem.

468. Education is a great tool. We want it to be *fast*, cheap, and pleasant. Which of these is <u>*not*</u> cheap?

A) Reading books
B) Watching movies, television, and videos
C) Going to college to hear lectures

A year at the top-ranked universities cost around $50,000 nowadays. (The price of tuition has been climbing faster than inflation for many years now. By the time you read this, $50,000 may seem cheap.)

$50,000 buys you two semesters of lectures.

A semester is about 15 weeks = 75 days. (15 × 5 = 75)

A day is 3 hours of lecture. (A normal load of courses is 15 units. That means 15 hours of lectures per week. That means 3 hours of lecture per day.)

Using conversion factors . . .

$$\dfrac{\$50{,}000}{1\ \text{year}} \times \dfrac{1\ \text{year}}{2\ \text{semesters}} \times \dfrac{1\ \text{semester}}{75\ \text{days}} \times \dfrac{1\ \text{day}}{3\ \text{hours of lecture}}$$

= **$111 per lecture hour.**

And this isn't paying some guy $111/hour to tutor you individually. This is group instruction. Some college classes have over 200 students in them.

This is insanely expensive.

150

470. Which of these are commutative?

{ shutting your eyes
  going to sleep   Not commutative. It would be hard to fall asleep and
                          then close your eyes.

{ dreaming
  waking up   Not commutative. It is much easier to dream before
                          you wake up.

{ dreaming about locomotives
  dreaming about pizza   These could be done in either order. They are
                          commutative.

471. $0.07\overline{)28.63}$ $7\overline{)2863.}$ with quotient 409.

Twelve is what percent of 18?   12 = ?% of 18 $12 \div 18 = \frac{2}{3} = 66\frac{2}{3}\%$

What is 37½% more than 24?

$37\frac{1}{2}\% = \frac{3}{8}$ $\frac{3}{8}$ of 24 is 9.   9 + 24 = 33

472. They saw each other at the end of a 400-foot hallway. They walked toward each other. Lucy was walking 4 feet per second faster than Grandma Amy. In 25 seconds they met and hugged.

How fast was Grandma Amy walking?

The first box to fill in is the one describing what we are looking for.

	d	r	t
Grandma Amy		r	
Lucy			

	d	r	t
Grandma Amy		r	25
Lucy		r + 4	25

	d	r	t
Grandma Amy	$25r$	r	25
Lucy	$25(r + 4)$	$r + 4$	25

We know that the total distance covered by both of them together is 400 feet.

$$25r + 25(r + 4) = 400$$

Distributive law $25r + 25r + 100 = 400$

Combine like terms $50r + 100 = 400$

Subtract 100 from both sides $50r = 300$

Divide both sides by 50 $r = 6$

Grandma Amy was walking at the rate of 6 feet per second.

473. Lucy was now 7. How many hours will it be before she turns 16? (Use 365 days = 1 year.)

We want to convert 9 years $(16 - 7 = 9)$ into hours.

$$\frac{9 \text{ years}}{1} \times \frac{365 \text{ days}}{1 \text{ year}} \times \frac{24 \text{ hours}}{1 \text{ day}}$$

$$= \frac{9 \text{ years}}{1} \times \frac{365 \text{ days}}{1 \text{ year}} \times \frac{24 \text{ hours}}{1 \text{ day}} = 78,840 \text{ hours}$$

474. Lammy told Lucy that he lend 70% of the value of the station. Lammy's appraiser said the station was worth $360,000. How much would the loan be?

70% of $360,000
0.7 × 360,000
$252,000

475. Joe might get 2% of that 1,000 **HaPPiNeSS PoiNtS** if he were thinking about creating a railroad. How many **HaPPiNeSS PoiNtS** is that?

2% of 1,000 $= 0.02 \times 1,000 = 20$ **HaPPiNeSS PoiNtS**. Joe gets more pleasure (30 **HaPPiNeSS PoiNtS**) just eating popcorn.

479. In an hour the receptionist could answer 8 phone calls or seat 11 parties. Luke could answer 5 phone calls or seat 7 parties. Who should do what?

Production chart			**Opportunity chart**	
	phone calls	seating	phone calls	seating
receptionist	8	11	$\frac{11}{8}$	$\frac{8}{11}$
Luke	5	7	$\frac{7}{5}$	$\frac{5}{7}$

One of the easiest ways to compare the fraction in the opportunity chart is to approximate each of them as a decimal. Using a calculator, type in 11 and then type in ÷ and then type in 8 and then =, and get 1.375.

Doing that for all four fractions, we get . . .   **Opportunity chart**

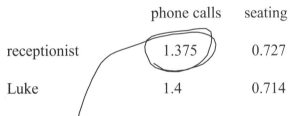

	phone calls	seating
receptionist	1.375	0.727
Luke	1.4	0.714

The receptionist has a smaller opportunity cost for phone calls, and she should do that.

Luke has a smaller opportunity cost for seating, and he should do that.

480. When Lucy plans and negotiates, she makes (on the average) $2,000 per hour. If she paints walls, she makes $37 per hour. If she sweeps floors, she makes $9 per hour.

What is her opportunity cost if she paints walls?

Opportunity cost is defined to be *the value of the best thing that you are giving up.* If Lucy is busy painting walls, she is giving up doing planning and negotiating (*Nice parallelism! —Mrs. Grumkey, your eighth grade teacher*) and her opportunity cost is $2,000 per hour.

486. If x is an odd number, what is the next consecutive odd number?

If x were, for example, 9, then the next consecutive odd number would be 11.

If x were, for example, 83, then the next consecutive odd number would be 85.

The next consecutive odd number after x would be x + 2.

487. Use the "six pretty boxes" approach on the previous problem to find the equation for:

Lucy's train from Reno to Sacramento will travel at 30 mph. When it goes back from Sacramento to Reno it will go 20 mph since it will be traveling uphill and will take 2 hours longer. How long will it take to go from Reno to Sacramento?

The first box we fill in is the thing we want to find out.

	d	r	t
going from Reno to Sacramento			t
going from Sacramento to Reno			

You can use any variable you like: $a, b, c, d, e, \ldots, w, x, y,$ or z. I like to use t when time is involved.

Since it will take two hours longer going the other direction:

	d	r	t
going from Reno to Sacramento			t
going from Sacramento to Reno			$t + 2$

We are told the rates, r, in each direction:

	d	r	t
going from Reno to Sacramento		30	t
going from Sacramento to Reno		20	$t + 2$

154

Knowing distance = rate times time (d = rt), we can fill in the last two boxes:

	d	r	t
going from Reno to Sacramento	$30t$	30	t
going from Sacramento to Reno	$20(t + 2)$	20	$t + 2$

Usually, it is the last two boxes that you fill in that will give you the equation. In this case, we know that the distance from Reno to Sacramento is the same as the distance going the other way.

$$30t = 20(t + 2)$$

We will be able to solve this equation in the next chapter after we learn the distributive law: $a(b + c) = ab + ac$.

488. Lucy could redecorate 12 locomotives every 25 seconds. How long would it take her to redecorate all 126 locomotives in her cereal bowl?

We want to convert 126 locomotives into seconds. We know that 12 locomotives matches up with 25 seconds. The conversion factor will be either

$$\frac{12 \text{ locomotives}}{25 \text{ seconds}} \quad \text{or} \quad \frac{25 \text{ seconds}}{12 \text{ locomotives}}$$

$$\frac{126 \text{ locomotives}}{1} \times \frac{25 \text{ seconds}}{12 \text{ locomotives}}$$

$$= 262.5 \text{ seconds}$$

489. Lucy had a dream about one of her trains carrying 176 people. She knew that each coach car could carry 24 passengers. The train would need 8 train workers. How many coach cars would there be?

Let x = the number of coach cars.
Then 24x = the number of passengers on the train.
Then 24x + 8 = the number of people on the train.

$$24x + 8 = 176$$

Subtract 8 from both sides $\quad 24x = 168$

Divide both sides by 24 $\quad x = 7$

There would be 7 coach cars on that train.

490. Express $\frac{3}{8}$ as a percent.

If you have memorized the Nine Conversions (in *Life of Fred: Decimals and Percents*), you have it easy. It is 37½%.

Otherwise, you have to divide it out:

$$
\begin{array}{r}
0.375 \\
8\overline{)3.000} \\
\underline{24} \\
60 \\
\underline{56} \\
40 \\
\underline{40} \\
\end{array}
$$

½ = 50%	¼ = 25%	¾ = 75%
⅓ = 33⅓ %	⅔ = 66⅔%	⅛ = 12½%
⅜ = 37½%	⅝ = 62½%	⅞ = 87½%

491. cat

492. The very same locomotive was offered by two different companies.
Coalback Loco—$50,000. We are located in Kansas.
Freedonia Engines—$30,000 and we offer free shipping to Kansas.

The choice was easy, except that the United States put a tax (a tarriff) on locomotives purchased from companies in other countries. It was a $40,000 tax. Who benefits from this tariff?

The only one who benefits is **Coalback Loco**.

493. Lucy estimated 20 would choose ribs, 30, spaghetti, and 50, chow mein. Draw a bar graph illustrating this.
Chow mein (mein is pronounced like the state of Maine) is stir-fried vegetables with meat—usually chicken, scrimp, or beef—served with fried noodles.

494. Who should be doing dishes? (Translation: Who has the smaller opportunity cost?)

Simplifying the fractions . . .

we see that Lucy's mom has a smaller opportunity cost than Luke. 0.06 < 0.1 She should do the dishes (and Luke make the beds).

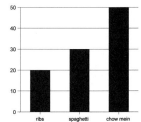

Opportunity Cost Chart

	🍽	🛏
Lucy's mom	0.06	$\frac{200}{12}$
Luke	0.1	$\frac{40}{4}$

511. If Lucy is charging 5¢ per foot to transport Scizzzors across her bedroom floor, how much would it cost to move Scizzzors 3 meters? (Your conversion factor will be based on 1 feet ≐ 0.3048 meters.)

The conversion factor will be either $\dfrac{1 \text{ foot}}{0.3048 \text{ m}}$ or $\dfrac{0.3048 \text{ m}}{1 \text{ foot}}$

$$\dfrac{5 ¢}{1 \text{ foot}} \times \dfrac{1 \text{ foot}}{0.3048 \text{ m}}$$

$$= \dfrac{5 ¢}{1 \text{ foot}} \times \dfrac{1 \text{ foot}}{0.3048 \text{ m}}$$

$$\doteq \dfrac{16.404 ¢}{\text{meter}}$$

And since we want to go 3 meters, the cost will be 3 × 16.404¢ ≐ 49¢.

(≐ means "equals after rounding off.")

512. Lucy didn't like the way Scizzzors looked. She thought that crocodiles should have tails. She got a paperclip and attached it to the

doll. That increased its weight by 5% Now she thought he looked much more like a real crocodile. How much did Scizzzors now weigh?

There are two ways to do a "5% more" problem.

The longer way:
　　5% of 864 is 43.2　　　　　(0.05 × 864 = 43.2)
　　Add 43.2 grams to the original 864 grams. He now weighs 907.2 grams.

The shorter way:
　　5% more than the original 100% is 105%. (You can always do that step in your head.)
　　105% of 864 = 1.05 × 864 = 907.2 grams

The longer way is one multiplication and one addition.
The shorter way is one multiplication.

*　　If you do a hundred of these problems and you save 5 seconds on each of them, you'll save 500 seconds (8⅓ minutes), which is enough time to enjoy a small pizza.*

513. Lucy's allowance was $1.25 per week. How long would it take for her to pay off the $10,000 purchase?

We want to convert $10,000 into weeks.

$$\frac{\$10{,}000}{1} \times \frac{1 \text{ week}}{\$1.25} = 8{,}000 \text{ weeks}$$

For fun, let me convert that to years.

$$\frac{8000 \text{ weeks}}{1} \times \frac{1 \text{ year}}{52 \text{ weeks}} \doteq 154 \text{ years}$$

514. Education is a great tool. We want it to be *fast*, cheap, and pleasant. Which of these is *fast*?

A) Reading books
B) Watching movies, television, and videos
C) Going to college to hear lectures

If you watch movies, television, and videos, the rate that information is transmitted is really slow. The stuff that you learn in a two-hour movie can be presented in twenty minutes of reading.

If you attend college lectures, you have to get dressed, go to the classroom, listen to a professor who is sometimes not well prepared (Fred is the exception), watch them scribble on the blackboard, take notes, and head back to your house. 60–80 minutes expended.

Reading is the *fast* road to education. Walk seven steps from the kitchen table to your desk. No need to get dressed. The presentation has been carefully prepared. (It takes me about an hour per page to write *Life of Fred* books.) The notes have already been done since it's all written down for you. Three hours of reading per day is probably equivalent to six hours of college lectures or zillions of hours of video watching.

Summary: If you want pleasant, watch the videos. If you are rich and want the pretty life, then go to college. If you want to get it done—if you have a goal—books are the way to go.

515. If you are writing a news report about Lucy being a kindergarten dropout, there are famous question words every reporter uses. Five of them are: When, Who, How, Why, and Where—4 W's and one H.

What is the sixth famous question word?

What is the sixth famous question word.

516. The monthly service fee is $64.48 plus 52¢ per call. In the first month the bill was $130. How many calls had Lucy made?

Let x = the number of calls that Lucy had made.
Then $52x$ = the cost of the calls (in pennies).
Then $52x + 6448$ = the total cost of the bill (in pennies).

$$52x + 6448 = 13000$$

Subtract 6448 from both sides $\quad 52x = 6552$
Divide both sides by 52 $\quad\quad\quad\quad x = 126$

Lucy had made 126 calls.

Or we could have done the problem in dollars instead of cents.

Let x = the number of calls that Lucy had made.
Then $0.52x$ = the cost of the calls.
Then $0.52x + 64.48$ = the total cost of the bill.

$$0.52x + 64.48 = 130$$

Subtract 64.48 from both sides $\quad 0.52x = 64.52$
Divide both sides by 0.52 $\quad\quad\quad\quad x = 126$

517. If two consecutive even numbers add to 1,294, what are those numbers?

Let x = the first consecutive even number.
Then $x + 2$ = the next consecutive even number.

$$x + x + 2 = 1294$$

Combining like terms $\quad\quad 2x + 2 = 1294$
Subtract 2 from both sides $\quad 2x = 1292$
Divide both sides by 2 $\quad\quad\quad x = 646$

The two consecutive even numbers that add to 1,294 are 646 and 648.

If you had just written 646 as the answer, you would not have answered the question of what are those numbers.

540. Under the old system (8 hours sleep): *x ideas/hour in 3 hours.*
Under the new system (10 hours sleep): *x + 4 ideas/hour in 2 hours.*

Both systems produced the same number of ideas. Lucy was delighted. How many ideas per hour did she get under the old system?

	d ideas produced	r ideas per hour	t hours
old system 8 hours sleep		x	3
new system 10 hours sleep		$x + 4$	2

	d ideas produced	r ideas per hour	t hours
old system 8 hours sleep	$3x$	x	3
new system 10 hours sleep	$2(x + 4)$	$x + 4$	2

Since both systems produced the same number of ideas . . .

$$3x = 2(x + 4)$$
$$3x = 2x + 8$$
$$x = 8$$

Under the old system of 8 hours sleep, Lucy was producing 8 ideas/hour.

541. One $5 bill fell off the van. What percent of the $400,000 was that?

$5 is what percent of $400,000?
5 = ?% of 400,000

If you don't know both sides of the *of,* divide the number closest to the *of* into the other number.

$$5 \div 400,000 = 0.0000125 = 0.00125\%$$

Or without a calculator . . .

$$
\begin{array}{r}
0.0000125 \\
400,000\overline{)\,5.0000000} \\
\underline{400000} \\
1000000 \\
\underline{800000} \\
2000000 \\
\underline{2000000} \\
\end{array}
$$

160

542. One gallon of perfume that Lucy received for her birthday cost $2.07. How much would a liter of that stuff cost?

One gallon ≈ 3.785 liters.

We want to convert 1 liter into a cost.
We know two things: ① 1 gallon cost $2.07 and
② 1 gallon ≈ 3.785 liters. That gives us two conversion factors.

$$\frac{1 \text{ liter}}{1} \times \frac{1 \text{ gallon}}{3.785 \text{ liters}} \times \frac{\$2.07}{1 \text{ gallon}}$$

$$= \frac{1 \text{ liter}}{1} \times \frac{1 \text{ gallon}}{3.785 \text{ liters}} \times \frac{\$2.07}{1 \text{ gallon}}$$

≈ $0.5468956 ≐ $0.55   or 55¢   (I used my calculator.)

543. Luke had 9 pennies and a 4.6-gram ball of gum in his left pocket. That weighed the same as the 6 pennies and the 16-gram rock he had in his right pocket.

How much does a penny weigh?

Let x = the weight of a penny.
Then 9x + 4.6 = the weight in his left pocket.
Then 6x + 16 = the weight in his right pocket.

$$9x + 4.6 = 6x + 16$$

Subtract 6x from both sides $3x + 4.6 = 16$
Subtract 4.6 from both sides $3x = 11.4$
Divide both sides by 3 $x = 3.8$

Each penny weighed 3.8 grams.

544. She realized that she didn't need 16 phone lines. Two lines would be plenty for her right now. What percent decrease in the number of phone lines would that be?

To go from 16 down to 2 is a decrease of 14.
14 is what percent of 16?
14 ÷ 16 = 0.875 = 87.5% decrease. This could also be written as 87½%.

545. If x is an odd number, what are the next two consecutive odd numbers?

The next two consecutive odd numbers after x are x + 2 and x + 4. For example, if x were 35, then x + 2 would be 37 and x + 4 would be 39.

546. Let the domain be the set of students at KITTENS and the codomain be the set of all humans. Mike and Ike are students at KITTENS. They are twins. Make an argument why *Associate to each student their biological mother* is still a function.

Again, we have to read the definition carefully: A function is any rule that assigns to each element of the domain exactly one element of the codomain. It does not say that elements of the first set are assigned to distinct elements in the codomain. All it says is that each element of the domain has one "answer" in the codomain.

Addition is a function. It assigned to each ordered pair of numbers an answer. (3, 5) is assigned to 8. (2, 533) is assigned to 535.

Give me an ordered pair and there will always be exactly one answer. ← definition of function

The fact to (3, 5) is assigned to 8 and (1, 7) is also assigned to 8 doesn't stop this from being a function.

547.
$$6y + 59 = 8y + 10 + 5y$$

Combine like terms	$6y + 59 = 13y + 10$
Subtract 6y from both sides	$59 = 7y + 10$
Subtract 10 from both sides	$49 = 7y$
Divide both sides by 9	$7 = y$

548. Is this a function from {7, 8} to {#, @, 5}?

$$7 \rightarrow \#$$
$$8 \rightarrow \#$$

Yes. Each element in the domain is assigned to exactly one element in the codomain.

We note that this function doesn't happen to be one-to-one. Not all functions are one-to-one.

549. If the legs of a right triangle are 6 and 8, what is the length of the longest side?

The Pythagorean theorem ⇨

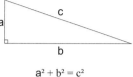

$$a^2 + b^2 = c^2$$

In this problem a will be 6 and b will be 8.

$$6^2 + 8^2 = c^2$$

arithmetic $\qquad 36 + 64 = c^2$

more arithmetic $\qquad 100 = c^2$

What number times itself equals 100? That's easy. c = 10.

550. In most walks of life there are two subjects that are needed above all the rest. The most important is English.

Stop! You can't say that! This is a math book. I, your reader, expect you, Mr. Author, to support math. You say english!

The word *English* is capitalized.

Don't interupt!

It's interrupt.

And what do you mean by "English"?

I mean the ability to communicate in that language: speaking, reading, and writing.*

❋ Speaking without saying *um* or *you know*.

❋ Reading and being able to learn by reading. Many of the college-entrance tests give you a paragraph to read and then ask you questions about what that paragraph meant.

❋ Writing to ✓inform

✓entertain or

✓convince

without tons of grammatical errors.

In second place is mathematics. Even in science and engineering, if the person can't communicate effectively, their ability to do ⅓ ÷ ⅝ isn't of much value.

---

* If this were a German textbook, I would be saying the same thing about speaking, reading, and writing in German.

555. This is the production chart for Lucy, her receptionist, and Luke showing how much painting each can do (in cans of paint per hour).

Lucy	3
receptionist	2
Luke	1.5

The doorbell rings. Which one should go answer the door?

Luke should answer the door. The least amount of work will be lost if he stops working.

556. The past tense of *sing* is *sang*. Today I sing. Yesterday I sang. The past tense of *smile* is *smiled*. Today I smile. Yesterday I smiled. What is the correct spelling of the past tense of *panic*?

Verbs ending in *c* have a very weird past tense.

panic ➡ *panicked*
mimic ➡ *mimicked*
picnic ➡ *picnicked*
traffic ➡ *trafficked*

To traffic means to trade or deal in a specific thing, often in illegal things. *Today I traffic in heroin. Yesterday I trafficked in heroin.* Tomorrow I sit in jail.

557. Lucy found that after her morning nap she could write 7 more emails per hour than before her nap.

In the 2 hours before her nap and the 2 hours after her nap, she wrote a total of 158 emails. What was her rate of writing emails before her nap?

There are two approaches to this problem: ① Let x = . . . and ② six pretty boxes. I'm going to do it both ways. You can chose your favorite.

Let r = the emails per hour before her nap.
Then r + 7 = the emails per hour after her nap.

Complete Solutions and Answers

Then 2r = emails produced in two hours before her nap.
Then 2(r + 7) = emails produced in two hours after her nap.
Then 2r + 2(r + 7) = the total emails produced.

$$2r + 2(r + 7) \;=\; 158$$

Distributive law	$2r + 2r + 14 \;=\; 158$
Combine like terms	$4r + 14 \;=\; 158$
Subtract 14 from both sides	$4r \;=\; 144$
Divide both sides by 4	$r \;=\; 36$

Lucy was writing 36 emails per hour before her nap.

Now with six pretty boxes . . .

	d emails produced	r emails written per hour	t hours
before nap		r	2
after nap		$r + 7$	2

	d emails produced	r emails written per hour	t hours
before nap	$2r$	r	2
after nap	$2(r + 7)$	$r + 7$	2

$$2r + 2(r + 7) \;=\; 158$$

Distributive law	$2r + 2r + 14 \;=\; 158$
Combine like terms	$4r + 14 \;=\; 158$
Subtract 14 from both sides	$4r \;=\; 144$
Divide both sides by 4	$r \;=\; 36$

570. On Lucy's last hike she took the north route for t hours, drinking 2.2 cups of water per hour.

     Coming back home on the south route she took two hours longer and drank 1.2 cups of water per hour.

     She drank the same amount of water on each route. How long did she walk on the north route? (Translation: Find t.)

	d cups	r cups per hour	t number of hours
north route		2.2	t
south route		1.2	$t + 2$

	d cups	r cups per hour	t number of hours
north route	$2.2t$	2.2	t
south route	$1.2(t + 2)$	1.2	$t + 2$

We know that Lucy drank the same amount of water on each route.

$$2.2t = 1.2(t + 2)$$

Distributive law $2.2t = 1.2t + 2.4$

Subtract 1.2t from both sides $t = 2.4$ hours on the north route

571. Is *sixth* a cardinal number? Is *six* a cardinal number?

     Sixth is an ordinal number. The ordinal numbers are first, second, third, fourth, and so on.

     Six is a cardinal number. You use cardinal numbers to count the members of a set. If we let T = the set of those first six diapers, then, in symbols, card T = 6.

166

580. Lucy's blood pressure had been 120/80. It went up 25% upon hearing the good news. What was her current blood pressure?

25% more than 120 is 125% of 120, which is 1.25 × 120, which is 150.
25% more than 80 is 125% of 80, which is 1.25 × 80, which is 100.
    Her new blood pressure is 150/100.

581. With his bike loaded, Luke took t seconds to get from his house to the station. He rode at 20 feet/second.

    On the way back home he could go 25 feet/second. The trip home took 72 seconds less than the trip to the station.

    Using six pretty boxes, find out how long it took Luke to get from home to the station.

	d	r	t
toward the station		20	t
toward home		25	t - 72

	d	r	t
toward the station	20t	20	t
toward home	25(t - 72)	25	t - 72

Since the distance from home to the station is the same as the distance from the station back to home . . .

$$20t = 25(t - 72)$$

Distributive law	$20t = 25t - 1800$
Add 1800 to both sides	$20t + 1800 = 25t$
Subtract 20t from both sides	$1800 = 5t$
Divide both sides by 5	$360 = t$

It took Luke 360 seconds (which is 6 minutes) to get to the station.

167

582. Lucy slept for 8 hours. If she had slept 20% longer, how long would she have slept? Express your answer in hours and minutes.

There are two ways to do a "20% more" problem.

The long way:
20% of 8 is 0.2 × 8, which is 1.6   one multiplication
1.6 plus the original 8 equals 9.6 hours.   one addition

The shorter way:
20% plus the original 100% is 120%   one super easy addition
120% of 8 is 1.2 × 8, which is 9.6 hours   one multiplication

9.6 hours = 9 hours + 0.6 hours.
Converting 0.6 hours into minutes . . .

$$\frac{0.6 \text{ hours}}{1} \times \frac{60 \text{ minutes}}{1 \text{ hour}} = 36 \text{ minutes}$$

Lucy would have slept for 9 hours and 36 minutes.

583. Solve $6x - 13 = 2x + 9$

$$6x - 13 = 2x + 9$$
Subtract 2x from both sides $4x - 13 = 9$
Add 13 to both sides $4x = 22$
Divide both sides by 4 $x = \dfrac{22}{4} = 5\frac{1}{2}$

or 5.5 if you like decimals

584. How many ordered ribs but didn't pay in cash?

So far we know ☞

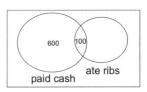

We know that 400 ate ribs,
so 300 (= 400 – 100) ate ribs but didn't pay in cash.

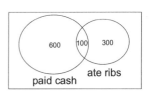

168

600. He ordered 5 pounds of popcorn. In terms of pickup truckloads, how much was that? (One ounce of popcorn = 2,000 cubic inches. One pound = 16 ounces. One cubic foot = 1728 cubic inches. One cubic yard = 27 cubic feet. A pickup truck can hold 3 cubic yards.)

We want to convert 5 pounds into truckloads.

$$\frac{5 \text{ lbs. of popcorn}}{1} \times \frac{16 \text{ ounces}}{1 \text{ lb.}} = 80 \text{ ounces of popcorn}$$

$$\frac{80 \text{ ounces of popcorn}}{1} \times \frac{2,000 \text{ cubic inches}}{1 \text{ ounce of popcorn}} = 160,000 \text{ in}^3$$

$$\frac{160,000 \text{ in}^3}{1} \times \frac{1 \text{ cubic foot}}{1728 \text{ in}^3} \doteq 92.592 \text{ ft}^3$$

$$\frac{92.592 \text{ ft}^3}{1} \times \frac{1 \text{ cubic yard}}{27 \text{ ft}^3} \doteq 3.429 \text{ cubic yards}$$

$$\frac{3.429 \text{ cubic yards}}{1} \times \frac{1 \text{ pickup truckload}}{3 \text{ cubic yards}} \doteq 1.14 \text{ truckloads}$$

601. Lucy's parents were thinking of adding another bedroom to their 1600 square foot house. It would be 10' × 12'. (ten feet by twelve feet)

What percent increase would that be?

A 10' × 12' room has an area of 120 square feet.

120 is what percent of 1600?

120 = ?% of 1600

If you don't know both sides of the *of*, you divide the number closest to the *of* into the other number.

120 ÷ 1600 = 0.075 = 7.5%

The addition of the extra bedroom would increase the size of the house by 7.5%.

622. $\frac{7}{8}$ is how much larger than $\frac{3}{5}$?

If you don't know whether to add, subtract, multiply, or divide, we use the **General Rule** that says, "Substitute easy numbers and notice which operation you use."

If we want to find out how much larger 6 is than 4, we know the answer is 2. We subtracted.

$$\frac{7}{8} - \frac{3}{5}$$

$$= \frac{35}{40} - \frac{24}{40}$$

multiplying the top and bottom of the first fraction by 5
multiplying the top and bottom of the second fraction by 8

$$= \frac{11}{40}$$

623. 13 issues of *Locomotive Monthly* and 31¢ were worth the same as 4 issues and $13. How much was each issue worth?

$+ \$0.31$

$+ \$13$

Let x = the price of one issue.

$$13x + 0.31 = 4x + 13$$

Subtract 4x from both sides $\quad 9x + 0.31 = 13$

Subtract 0.31 from both sides $\quad\quad 9x = 12.69$

Divide both sides by 9 $\quad\quad\quad x = 1.41$

The price of a single issue is $1.41.

624. Suppose some shoe is manufactured in sizes 8, 8½, 9, 9½, 10, 10½, 11, 11½, and 12. Is this discrete or continuous?

If it were continuous, then there would have to be a size 9⅛ (since there are sizes 9 and 9½). These shoe sizes are discrete.

625. One copy of Adventure Duck comic book costs 36¢.

When Luke bought 32 copies, they gave him a 16⅔% discount. (You remember that 16⅔% is the same as $\frac{1}{6}$ don't you?)

How much did he pay for a each comic book? Work in fractions rather than decimals.

If there was a $\frac{1}{6}$ discount, then Luke paid $\frac{5}{6}$ of the regular price.

$$\frac{5}{6} \text{ of } 36¢ = \frac{5}{6} \times \frac{36}{1} = \frac{5}{\cancel{6}_1} \times \frac{\cancel{36}^6}{1} = 30¢$$

626. Is this a function?

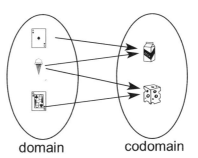

domain   codomain

If it is a function, then each member of the domain must be assigned to exactly one member of the codomain. But 🍦 is assigned to two different members of the codomain. It is not a function.

627. One company offered: 12 cans + $50 shipping.
A second company offered: 20 cans + free shipping

Twelve cans + $50 shipping from the first company would cost the same as 20 cans from the second company. Find out what that price per can is.

Let c = the price of one can of roof coating.
Then 12c + 50 cost the same as 20c.

$$12c + 50 = 20c$$

Subtract 12c from both sides $\quad\quad 50 = 8c$

Divide both sides by 8 $\quad\quad 6.25 = c$

A can of roof coating cost $6.25.

171

630. Lucy had had 0 pieces of real estate. She now had 1 piece of real estate. Was that more or less than 1,000% increase in her real estate holdings?

First, let's look at what 1,000% increase of, say, 6 means?
1000% more than 6 means 1100% (= 100 + 1000) of 6.
1100% of 6
11.00×6
66

Now, let's look at 1,000% increase of 0.
1000% more than 0 means 1100% of 0 = 11×0 = 0.

Any multiple of zero is zero.

Going from 0 pieces of real estate to 1 piece of real estate is more than a zillion percent increase.

631. **Coalback Loco** will sell it for $50,000.
Freedonia Engines will sell the same thing for $30,000, but there is an extra $40,000 tax because *Freedonia Engines* isn't local.

Who is hurt by this tariff?

Everyone is hurt except **Coalback Loco**. Everyone. The men and women who work at *Freedonia Engines* suffer. Lucy suffers.

Everyone who owns a railroad in the United States and must buy their engines at a higher price—they all must charge their customers more in order to stay in business. So all the customers are hurt. Virtually everyone in the United States owns things that have been delivered by train.

Why would Congress pass such an evil law?

We know that a law that put a special tax on things made by Jews, women, or racial minorities would be unjust. Why would it be acceptable to tax things made by someone who lives in a different city (or country) than you do?

Every human should be treated equally under the laws we pass.

172

632. One can serves 2. Three cans for 49¢. Lucy needed to be able to serve 100 people. Using conversion factors, how much is this going to cost Lucy?

We want to convert 100 people into a cost.

$$\frac{100 \text{ people}}{1} \times \frac{1 \text{ can}}{2 \text{ people}} = 50 \text{ cans}$$

$$\frac{50 \text{ cans}}{1} \times \frac{49¢}{3 \text{ cans}} \doteq 817¢ \quad \text{or} \quad \$8.17$$

633. How many of those 1,900 people didn't order ribs or pay in cash?

So far we know ☞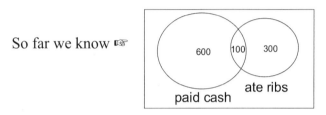

There are 1,000 people who either paid cash, ate ribs, or both. That leaves 900 people (= 1,900 – 1,000) who did none of these.

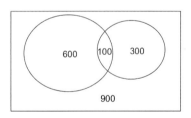

634. Let T be the set {hammer, screwdriver, saw}. Let L be the set of the 25 trips that Luke took to the station {#1, #2, #3, . . . , #24, #25}.

hammer → #1   screwdriver → #1   saw → #2

Is this a function from T to L? Is it one-to-one?

It is a function. Each member of T is assigned exactly one member of L. Translation: The hammer, screwdriver, and saw were all taken on some trip, and the hammer, for example, wasn't taken on two different trips.

It is not 1–1. Both the hammer and the screwdriver were taken on the same trip.

646. Find three errors in this letter.

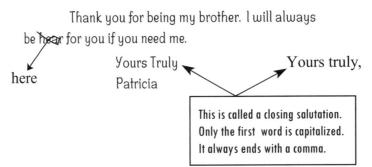

Dear Luke.

    Thank you for being my brother. I will always be ~~hear~~ for you if you need me.

here

Yours Truly
Patricia

Yours truly,

This is called a closing salutation.
Only the first word is capitalized.
It always ends with a comma.

647. Lucy's monthly storage bill, which was $500 (= 4 × $125), now became $495. What percent decrease was that?

    Going from $500 to $495 was a savings of $5.
    $5 is what percent of $500?
    $5 = ?\%$ of 500
    $5 \div 500 = 0.01 = 1\%$ savings

648. On her hike she spotted six train stations, which she called A, B, C, D, E, and F. She assigned each of those train stations to one of three categories: ① I'd like to buy it; ② I might buy it; and ③ I would never buy it.

    If the domain = {A, B, C, D, E, F} and the codomain = {①, ②, ③}, will this function be 1–1?

    It's not possible. If each of those six train stations are assigned to one of those three possibilities, either ①, ②, or ③ is going to receive more than one assignment.

    For example, if she assigned A → ② and B → ③ and C → ①, then when it came time to assign D to either ①, ②, or ③, the function could no longer be 1–1.

    Translation: If 6 pigeons were going to sit on 3 nests, then at least one of the nests would be crowded.

649. Lucy thought for a moment: I could get 49¢ for moving Scizzzors three meters. How far would Lucy Railroad have to move that crocodile in order to make a million dollars?

 There are two ways to do this problem.

First way: Use a conversion factor. Start with a million dollars and convert that into meters traveled.

$$\frac{\$1,000,000}{1} \times \frac{3 \text{ meters}}{\$0.49}$$

$$= 6,122,448.9 \text{ meters}$$

Second way: Use a proportion. (A proportion is two fractions that are equal to each other: $\frac{a}{b} = \frac{c}{d}$)

 Let x = how far the Lucy Railroad would have to move that crocodile in order to make a million dollars.

$$\frac{\$0.49}{3 \text{ meters}} = \frac{\$1,000,000}{x \text{ meters}}$$

$$\frac{0.49}{3} = \frac{1,000,000}{x}$$

cross multiply $\qquad\qquad$ $0.49x = 3,000,000$

divide both sides by 0.49 $\qquad\quad$ $x = 6,122,448.9$

 6,122,448.9 meters is hard to picture. For fun I'll convert that into miles. I'll do the work. You can just watch. **I, your reader, like that.** One meter is approximately 0.0006214 miles. (I looked that up on the Internet.)

$$\frac{6,122,448.9 \text{ m}}{1} \times \frac{0.0006214 \text{ miles}}{1 \text{ m}}$$

$$\doteq 3,800 \text{ miles}$$

It's about 3,200 miles from San Francisco to Portland, Maine. It's about 600 miles from Portland, Maine to Erie, Pennsylvania.

 Lucy will make a million dollars hauling Scizzzors on that trip.

650. The sum of three consecutive odd numbers is equal to 1,647. What are these numbers?

Let x = the first number.
Then x + 2 = the second number.
Then x + 4 = the third number.

$$x + x + 2 + x + 4 = 1,647$$

Combine like terms $$3x + 6 = 1,647$$

Subtract 6 from both sides $$3x = 1641$$

Divide both sides by 3 $$x = 547$$

The three numbers are 547, 549, and 551.

651. During the first 24 months of Lucy's life, her mom had made r entries into Lucy's baby book each month. In the second 24 months of Lucy's life, she made entries at the rate of r – 6 per month.

There were 1,296 entries at the end of those 48 months. Using six pretty boxes, find the value of r.

	entries made	rate	months
during the first 24 months		r	
during the second 24 months			

	entries made	rate	months
during the first 24 months		r	
during the second 24 months		r - 6	

	entries made	rate	months
during the first 24 months		r	24
during the second 24 months		r - 6	24

	entries made	rate	months
during the first 24 months	24r	r	24
during the second 24 months	24(r - 6)	r - 6	24

Usually, the equation comes from the last two boxes that are filled in.

We know that a total of 1,296 entries were made.

$$24r + 24(r - 6) = 1{,}296$$

Distributive law $\quad 24r + 24r - 144 = 1{,}296$

Combine like terms $\quad 48r - 144 = 1{,}296$

Add 144 to both sides $\quad 48r = 1{,}440$

Divide both sides by 48 $\quad r = 30$

Lucy's mom was make 30 entries per month in Lucy's baby book during the first 24 months of Lucy's life.

652. If you have a $1.25/week allowance, how long could you make payments on a $125/month storage unit?

At the end of the first month you would have about $5. (= 4 × 1.25) You make a payment of $5 and you would still owe $120. You couldn't even afford the first month's payment.

At the end of the second month you would owe $245 (= 120 + 125) You make a payment of $5 and you would owe $240.

Each month you would get deeper and deeper into debt.

With that income you couldn't afford that storage unit.

653. When the Lucy Railroad gets large and has 175 employees in its home office, roughly how many of them will do 80% of the work?

The Pareto principle states that 20% of them will do 80% of the work. 20% of 175 = 0.2 × 175 = 35.

Thirty-five of those 175 will be doing 80% of the work.

654. I can think of only 24 words that do not contain *a, e, i, o,* or *u.*

Try to name at least two of them before you look at my list in the answer.

I just gave you two of them.

Here are the others that I could think of: *by, cry, dry, fly, fry, gypsy, hymn, lymph, lynch, Lynn, lynx, myth, ply, pry, pygmy, rhythm, shy, sky, spry, try, tryst,* and *why.*

655. Find the value of n so that $2^n = 30 + 2$.

That is the same as asking for a value of n so that $2^n = 32$.

Since $2 \times 2 \times 2 \times 2 \times 2 = 32$ that means that $2^5 = 32$. n = 5

656. The total cost of the three gas masks, including the $3.98 shipping charge, was $25.82. How much did each gas mask cost?

Let x = the cost of one mask. ⤺ You always start the solution of a word problem with this line.

Then 3x = the cost of 3 masks.
Then 3x + 3.98 was the total cost.

$$3x + 3.98 = 25.82$$

Subtract 3.98 from both sides
Divide both sides by 3

$$3x = 21.84$$
$$x = 7.28$$

Each mask cost $7.28.

In algebra books they will eventually stop writing the

By just looking at the equations it is easy to guess how they got from one line to the next.

660. How many ideas per hour under the new system?

Looking at those six pretty boxes of the previous problem . . .

	d ideas produced	r ideas per hour	t hours
old system 8 hours sleep	$3x$	x	3
new system 10 hours sleep	$2(x+4)$	$x+4$	2

and using the fact that we found that x = 8, we know that she produced 12 ideas per hour under the new system. (8 + 4 = 12)

661. The cost of the food is generally about 30% of the price on the menu.
If a serving of ribs would cost Lucy $2.88, what should the menu price be?

$2.88 is 30% of the menu price
2.88 = 30% of ?
If you don't know both sides of the *of*, you divide the number closest to the *of* into the other number.
$2.88 \div 0.3 = 9.6$
The menu price for a serving of ribs should be $9.60.

662. This is an opportunity chart for two choices that Lucy has.

	Lucy	
	Railroad	Lounge
Lucy		$\frac{3}{5}$

The $\frac{3}{5}$ came from a production chart in which the railroad could make $3,000,000 in a year and the lounge could make $5,000,000 a year—or was it that the railroad could make $5,000,000 a year and the lounge make $3,000,000 a year?

Railroad $3M and the Lounge $5M. Then for every $5M that the Lounge makes she loses $3M in railroad revenue. For every $1M that the Lounge makes she loses $$\frac{3}{5}$ M in railroad revenue.

670. Earlier this morning Lucy had been paying her brother Luke $8/hour. She raised his salary to $11. What percent increase was that?

She increased his salary by $3.

$3 is what percent of $8?

3 = ?% of 8

3 ÷ 8 = 37½%   This is one of the Nine Conversions you might have memorized.

$$\text{Otherwise} \Rightarrow 8\overline{)3.000} \quad \frac{0.375}{}$$

671. Luke had ordered a Fred doll. He called it an "action figure" so that his friends wouldn't make fun of him. Under Luke's direction, Fred would go on all kinds of adventures. Luke would give Fred a hug 4 times each hour.

Lucy's dad would give his Fred doll a hug 5 times each hour. He called it a "good luck squeeze" so that his fellow workers wouldn't make fun of him.

Lucy hugged her Fred doll 6 times each hour. All her girl friends were jealous that she had such a lovable doll and they didn't.

How long would it take these three people to give a total of 50 hugs?

	hugs given	rate	time
Luke	$4t$	4	t
Lucy's dad	$5t$	5	t
Lucy	$6t$	6	t

They gave a total of 50 hugs. $4t + 5t + 6t = 50$

Combine like terms $15t = 50$

Divide both sides by 15 $t = 3⅓$

It would take them 3⅓ hours (or 3 hours and 20 minutes).

672. Solve $6.2x + 17 = 2.4x + 22.7$

Subtract 2.4x from both sides $3.8x + 17 = 22.7$

Subtract 17 from both sides $3.8x = 5.7$

Divide both sides by 3.8 $x = 1.5$

180

700. Eleven ducks plus an $8 sack of duck food cost the same as five ducks and a $22.58 sack of duck food. How much does a duck cost?

Let d = the cost of a duck.

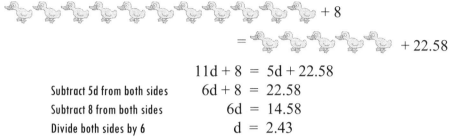

$$11d + 8 = 5d + 22.58$$

Subtract 5d from both sides $\quad 6d + 8 = 22.58$

Subtract 8 from both sides $\quad\quad 6d = 14.58$

Divide both sides by 6 $\quad\quad\quad d = 2.43$

A duck cost $2.43.

701. Create a function in which the domain is {5, 7} and the codomain is {egg, top}.

There are four possible answers you might have given.

First possible function   Second possible function   Third possible function   Fourth possible function

First	Second	Third	Fourth
5 → egg	5 → egg	5 → top	5 → top
7 → egg	7 → top	7 → egg	7 → top

702. Let T = the set of all train cars that Lucy is playing with. Let W = the whole numbers, which is {0, 1, 2, 3, 4, . . .}.

Is this a function from T to W? *Assign to each train car the number of wheels that it has.*

Yes. The first car might have 8 wheels. The second car, 10 wheels. The third car, 8 wheels, and so on.

The two parts of the definition of a function are true:

① *Each car* receives an assignment.

You pick up a car and look at it. It has a number of wheels that you can count. It will receive an assignment.

② Each car receives *exactly one* assignment.

Each car will have a definite number of wheels. It can't have both 6 wheels and 10 wheels.

703. If the domain is the thoughts of these three people {train set, statue removal, baby} and the codomain is {☺, ☹}, create a function and tell whether the function is one-to-one.

There are many possible functions you might have created. (Actually, there are 8 possible functions.)

One possible function: train set → ☺
                          statue removal → ☺
                          baby → ☺

A second possible function: train set → ☺
                               statue removal → ☹
                               baby → ☹

The most likely thing that will happen is that the baby will get the new bedroom and Lucy and Luke will be disappointed. That situation would be represented by the function: train set → ☹
                          statue removal → ☹
                          baby → ☺.

In any event, any function with three members in the domain and two members in the codomain can not be one-to-one. Translation: Two or more members of the domain will have the same image in the codomain.

704. The net monthly income from these stations was . . .
    $300 from station A
    –$517 from station B
    $889 from station C
    $1446 from station D
    –$266 from station E
    –$49 from station F.

Without using your calculator find the total income from these six stations.

The easiest way to add a bunch of positive and negative numbers together is add all the positive numbers in one pile and add all the negative numbers in another pile. Then combine those two answers.

300	–517	
889	–266	
1446	– 49	
2635	–832	$2635 + (–832) = \$1,803$

182

705. In economics the goal isn't to have everyone working 60 hours a week. It is to have abundant things. In life is the goal to have abundant things?

No.

Wait! Stop! I, your reader, can't believe that's the whole answer.

All the ads you see on television or see in magazines seem to say that the goal in life is to have lots of stuff. Any fool can see that just having lots of possessions is not the goal of life.

But why do you see so many people spending all their waking hours running around getting stuff?

They believe the ads.

Having enough to eat, some clothes, and a place to get out of the cold is the starting place. Then you want the safety of your health, your family, your property. Then loving and being loved. Then achievement and respect by others.

These are what Luther called, "Goods, fame, child and wife—let these all be gone." They are all nice, but they are nothing compared with what lies beyond them.

Okay. Tell me. What's beyond?

That would spoil your search. ☺

706. Lucy could have done the work herself that she was hiring Rammy, Wamford, and Hamlock to do. Why didn't she?

This is at the heart of Ricardo's Law of Comparative Advantage. Rammy's opportunity cost for doing the real estate work was less than Lucy's opportunity cost for doing that real estate work.

Wamford's opportunity cost for selling chairs and cups was less than Lucy's opportunity cost for doing that work.

Translation: Lucy could make more per hour doing those phone calls and thinking and planning for future I Love You Amy Stations than she could by doing real estate, buying restaurant supplies, or hiring people.

In contrast, Rammy's real estate business had been very slow. He had spent hours in his office talking with his buddies about how slow business was. His opportunity cost for doing work for Lucy was near zero.

183

730. Dad's roll of electrical wire had been 200 feet long. Lucy had used $162\frac{1}{4}$ feet of it to wire the outlets. How much was left on the roll?

$$200 \qquad\qquad 199 + \frac{4}{4}$$

$$-\;162\frac{1}{4} \qquad -\;162\frac{1}{4}$$

$$37\frac{3}{4} \text{ feet were left on the roll.}$$

731. Each morning it took her 3% longer to comb her hair than the previous morning. One June 1st it took her a minute to run a comb through her hair. How long would it be before she was spending two minutes combing her hair?

By the Rule of 72, it would take 72/3 = 24 days. (June 25th)

732. Beyond money, beyond fame, what is there to achieve?

After you have made your first million or your first billion, having more **MONEY** doesn't mean that much. You don't eat any better. Moving to a bigger mansion doesn't give you that much more of a thrill. There are only so many shoes that you can buy and wear.

After everyone knows your name, **FAME** can get to be a really bother. The President of the United States can't go shopping at the grocery store. He can't take his kids to the movies. A famous movie star can't go anywhere in public without a bunch of people crowding around saying, "Are you that movie star? Can I have your autograph?"

Famous people have to avoid the public.
My favorite Hollywood cowboy is Thomas Hezikiah Mix, known by most people as Tom Mix. He was called the "King of Cowboys."
What I like most about Tom Mix is that he was clean-cut. He was unflawed. He was wholesome. Not too many actors nowadays (zero?) can be described that way.

Complete Solutions and Answers

He made 291 movies and was a megastar. Millions of American kids would go to the movies each week to see him. Even his horse Tony was a celebrity.*

 Zillions of listeners heard the *Tom Mix Ralston Straight Shooters* radio series which was popular from most of the 1930s through the early 1950s. Those who listened were offered a series of 12 Ralston–Tom Mix Comic books. You had to mail in your request. Emails were not accepted.

You want **FAME**? Tom Mix had it. He was even commemorated on a United States Postage stamp.

Now to the question of what is beyond money and fame. Both of those are nice, but *is that all there is?*

Intermission

Many people—most people, in fact—don't even get as far as money and fame.

M & F almost always require dedication and hard work over years.

It's much more fun to hang out with your friends or play video games.

There are a hundred ways to answer the question of what lies beyond the goals of money and fame. All of the answers seem to involve *getting beyond yourself.*

Read the biographies of the truly great people. None of them went around singing the song of: I, I, I. Each of them lived for something bigger than getting money or notoriety.

✳ For Lucy it is the work—the creation of the Lucy Railroad. Her railroad will be beautiful.

---

* Now, if anyone ever asks you the name of Tom Mix's horse, you can tell them.

Complete Solutions and Answers

✳ For some it is the movement. Save the ducks. Promote free speech on campus. Eliminate smallpox from the world.

✳ For some it is the family. Create a home where spouse and kids can find tons of love.

✳ For some it is for Goodness' sake—as I, your author, express it on the dedication page of each of my books (page 5).

740. Fill in the missing entry in that opportunity chart.

The production chart is

	Train	Lounge
Lucy	3	5

For every \$3M makes in trains she could make \$5M in her restaurant.

	Train	Lounge
Lucy	$\frac{5}{3}$	$\frac{3}{5}$

Some readers may have already noticed that in opportunity charts, the two entries in the same row are **reciprocals** of each other.

The reciprocal of $\frac{1}{4}$ is $\frac{4}{1}$ and the reciprocal of $\frac{4}{15}$ is $\frac{15}{4}$.

In algebra, we will say that the reciprocal of $\frac{m}{h}$ is $\frac{h}{m}$.

741. In the first hour of Lucky Lucy's Lounge, each order of burritos gave a profit of \$21, and each order of chow mein gave a profit of \$14. There were 30 more orders for chow mein than for burritos. The total profit in that first hour from these two menu items was \$1,365.

How many burrito orders were there? Use six pretty boxes.

	d profit	r profit per order	t number of orders
burrito orders		21	x
chow mein orders		14	$x + 30$

	d profit	r profit per order	t number of orders
burrito orders	$21x$	21	x
chow mein orders	$14(x + 30)$	14	$x + 30$

$$21x + 14(x + 30) = 1{,}365$$
$$21x + 14x + 420 = 1{,}365$$
$$35x + 420 = 1{,}365$$
$$35x = 945$$
$$x = 27 \text{ orders of burritos}$$

770. Sixteen years of storage rental at $500/month (125 × 4) will cost a total of . . .

$$\frac{16 \text{ years}}{1} \times \frac{12 \text{ months}}{1 \text{ year}} \times \frac{\$500}{1 \text{ month}} = \$96,000$$

771. Luke cut out the rectangle of carpet that had the perfume spill on it and threw it out the window. Suddenly the living room smelled better. How many square inches had Luke cut out?

4 ft.

6 ft. 3 in.

4 feet = 48 inches (4 × 12 = 48)
6 feet = 72 inches
6 feet 3 inches = 75 inches

A = length times width = (75)(48) = 3,600 square inches.

772. If it takes 38 drops of sweat to learn how to do percents, and it takes 45% more sweat to learn about functions, how many drops will that be?

This is the "15% more" type of percent problems, as we called them in *Life of Fred: Decimals and Percents.* There are two ways to do them.

The harder way: 45% of 38 is 17.1.   (0.45 × 38 = 17.1)
17.1 plus the original 38 equals 55.1 drops of sweat.

The easier way: 45% plus the original 100% is 145%. (This you can always do in your head.)
145% of 38 is 55.1 drops of sweat.   (1.45 × 38 = 55.1)

773. What would stop Lucy from cutting down those three telephone poles and rolling them into the street?

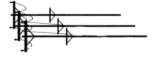

The first obvious difficulty is that four-year-olds would have a tough time cutting down telephone poles.

The second thing that would stop Lucy is the police. There has to be a law against throwing telephone poles in the street. It would be very easy to see where those poles came from—the stumps that were left in the ground.

The third possibility is that Lucy has grown up enough to know right from wrong.

774. A third company had a sale. They offered 19% off their regular price of $7.77 per can. Was that a better deal than the $6.25 per can that the first two companies offered?

If it is 19% off, then Lucy would be paying 81% of the regular price. 81% of $7.77 = $(0.81)(7.77) = 6.2937 \doteq 6.29 per can.

That is not a better price than the $6.25 that the first two companies offered.

775. On the day after Lucy's second birthday she learned her fourth word: *Toot!* Is *fourth* an ordinal or a cardinal number?

Fourth is an ordinal number. It describes the *order* in which things occur.

Lucy now knows four words. *Four* is a cardinal number.

776. Let's apply the Pareto principle to those 20%. This will say that 20% of those 20% will be doing 80% of the 80% of the work.

Simplify the math in the previous sentence.

20% of 20% will do 80% of the 80% of the work.
0.2×0.2 will do 0.8×0.8 of the work.
0.04 will do 0.64 of the work.
4% will do 64% of the work.

Find those four percent, fire the other 96%, and you'll still get more than half the work done.

Think of classical music composers. Five men—3 B's, a T, and an M—account for a substantial part of all classical music.

Great ability combined with great productivity is quite rare.

Naming the five "greats" in any field—physics, cooks, playwrights, piano players, mathematicians, religious leaders—and your list will be very similar to many other people's lists.

777. Six diaper changes per day. Each one take 5 minutes. Her mom is awake 16 hours per day. What percent of her day is spent changing diapers?

Six diaper changes at 5 minutes each will take 30 minutes.

30 minutes is what percent of 16 hours?

One-half hour is what percent of 16 hours?

$$\frac{1}{2} = ?\% \text{ of } 16$$

$$\frac{1}{2} \div 16$$ If you don't know both sides of the *of*, you divide the number closest to the *of* into the other number.

$$= \frac{1}{32}$$

$$= 0.03125 = 3.125\%$$

Lucy's mom spends approximately 3% of her waking hours working in the switching yard.

Switching Yards

Lucy thought a lot about switching yards. In her spare time she read magazine articles about railroads. In about a year from now when she would be 2, she would start to read books about railroads.

Switching yards are one kind of rail yard.

Rail yards are a complex bunch of railroad tracks used for sorting, loading, unloading, and storing railroad cars.[*]

Switching yards are where cars are separated and reassembled. For example, freight cars from all over the United States might be joined together for their trip to KITTENS University.

Lucy wanted to explain to her mother that changing her diaper was like getting a new caboose in the switching yard, but she couldn't. That's because the only word Lucy knew was "yucky."

---

[*] When Lucy gets a little older, she will be able to tell you about all the different rail yards: up yards, down yards, hump yards, gravity yards, flat yards, staging yards, transfer yards, receiving yards, sorting yards, and coach yards. Right now, all she knows about are switching yards.

780. Let D = the set of all 32 Adventure Duck comic books that Luke bought. Let T = the set of train books that Lucy got from the KITTEN library. Is this a function? *Assign to each member of D to Prof. Eldwood's* Locomotive Electrical Diagrams.

      Yes. Each comic book is assigned to exactly one member of the train books that Lucy got from the KITTENS library. That's the definition of function.

781. Water pouring in at 3 gallons per minute. Water draining out at 1.2 gallons per minute. (Lucy had accidently kicked the plug out.) How long until the 55-gallon tub is filled?

      It was filling at the rate of 1.8 gallons per minute. (3 − 1.2 = 1.8) Do we add, subtract, multiply, or divide. Using the 𝕲𝖊𝖓𝖊𝖗𝖆𝖑 𝕽𝖚𝖑𝖊, we use some simple numbers. To fill an 8-gallon tub with water flowing in at 2 gallons per minute will take 4 minutes. We divide.

      55 gallons ÷ 1.8 gallons per minute ≐ 30.55 minutes.

Converting 0.55 minutes into seconds using a conversion factor . . .

$$\frac{0.55 \text{ minutes}}{1} \times \frac{60 \text{ seconds}}{1 \text{ minute}} \doteq 33 \text{ seconds}$$

So in about 30 minutes and 33 seconds the tub will be full.

782. Lucy had dreams of creating Lucy Railroad. It would be bigger than any other railroad in the country.

      If the second biggest railroad had 6,308 cars, she knew she would have 250% more cars.

      How many cars would Lucy Railroad have?

      There are two ways to do "15% more" problems.

the harder way:
      250% more than 6,308 is 15,770.
      15,770 plus the original 6,308 is 22,078 cars in the Lucy Railroad

the easier way:
      250% plus the original 100% is 350%
      350% of 6,308 is 22,078 cars in the Lucy Railroad

783. Why does Congress pass laws restricting U.S. citizens from buying sugar from other countries?

    A) Members of Congress hate foreigners.

    B) Members of Congress hate dental cavities.

    C) Although sugar crops are only 2% of the total U.S. crop production, the sugar producers give 33% of the total crop industries' campaign contributions.

    It's $$$ in the pockets of Congressmen and Congresswomen. C)

784. Lucy listed all her habits and called that set H. She assigned each habit of hers into one of three categories: ① Those habits that she definitely wanted to keep; ② Those habits that neither helped nor harmed her; ③ Those habits she wanted to get rid of.

    She had 26 habits in her list (set H). Is this a function, and, secondly, is it one-to-one?

    Each element of H—such as her habit of nail-biting—was assigned to exactly one of the three categories. This is a function.

    Because there were 26 elements of the domain and only three elements in the codomain, at least two of her habits would have to be assigned to the same element in the codomain. It couldn't be 1–1.

785. Five-sixths of her ideas were worth nothing. One-sixth of them were worth $1,000 each. Under her new system, how much was she making per hour during her planning hours?

    Lucy was creating 12 new ideas per hour under her new system (where she got more sleep).

    One-sixth of those 12 ideas were valuable.

$$\frac{1}{6} \times 12 = 2 \text{ ideas per hour were valuable.}$$

$$\frac{2 \text{ good ideas}}{1 \text{ hour}} \times \frac{\$1,000}{1 \text{ good idea}}$$

$$= \frac{2 \text{ good ideas}}{1 \text{ hour}} \times \frac{\$1,000}{1 \text{ good idea}} = \$2,000 \text{ per hour}$$

Prof. Eldwood often points out that planning can be valuable.

800. Rammy located 96 vacant train stations in Kansas.

He reported to Lucy:

34 have air conditioning installed

35 have good roofs

43 have unbroken windows

7 have air conditioning and good roofs

10 have air conditioning and unbroken windows

9 have good roofs and unbroken windows

3 have air conditioning, good roofs and good windows

How many have no air conditioning, bad roofs, and broken windows?

Draw the
Venn diagram ☞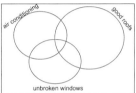

There is only one possible
number to stick in the
diagram.

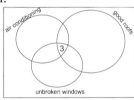

Once we have the "3" and we know that 7 have air conditioning and good roofs that gives us the 4 (= 7 – 3).

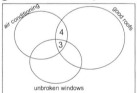

Similarly ☞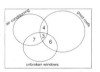

In the air conditioning circle
we have accounted for 14
(= 7 + 3 + 4) of them. That
leaves 20 (= 34 – 7 – 3 – 4).

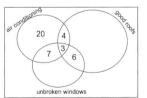

Similarly,

That leaves 7 (= 96 – 20 – 7 – 3 – 4 – 22 – 6 – 27).

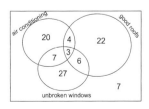

830. $5\frac{1}{5} - 2\frac{1}{3}$

$$
\begin{array}{cccc}
5\frac{1}{5} & 5\frac{3}{15} & 4\frac{15}{15}+\frac{3}{15} & 4\frac{18}{15} \\
-2\frac{1}{3} & 2\frac{5}{15} & 2\frac{5}{15} & 2\frac{5}{15} \\
\hline
& & & 2\frac{13}{15}
\end{array}
$$

831. The horsepower requirements are given by the formula: HP = 377n where n is the maximum number of cars to be pulled.

What is the horsepower requirements of a locomotive that will be pulling 4 cars?

HP = (377)(4) = 1,508.

832. Solve $\qquad$ $700x + 398 = 936x + 67.6$

Subtract 700x from both sides $\qquad$ $398 = 236x + 67.6$

Subtract 67.6 from both sides $\qquad$ $330.4 = 236x$

Divide both sides by 236 $\qquad$ $1.4 = x$

833. 3,600 square inches is how many square yards? Since there are 36 inches in a yard, there are 36^2 square inches in a square yard.

We want to convert 3,600 square inches into square yards. That sounds like a conversion factor problem. We know 1,296 square inches equals 1 square yard. $(36^2 = 1,296)$

$$
\frac{3,600 \text{ square inches}}{1} \times \frac{1 \text{ square yard}}{1,296 \text{ square inches}}
$$

≈ 2.777777 square yards

Cammy cut the 4 foot by 6 foot 3 inch rectangle. The carpet was $21 per square yard, so the cost was 2.777777 times $21 which was $58.33 (after rounding).

What Lucy's father failed to mention was that the original carpet was blue. The carpet that Cammy sent them was orange. For years they would look at that spot in the living room and remember the gallon of perfume that had given to their three-year-old for her birthday.

834. If a candy bar would normally cost \$2.20, how much would it cost with a 87% increase in price?

There are two ways to do a "87% more" problem.
The long way:

87% of 2.20 $= 0.87 \times 2.20 \doteq 1.91$

$2.20 + 1.91 = \$4.11$ would be the cost of that candy bar.

The shorter way:

87% + 100% $= 187\%$ (This part you can do in your head.)

187% of 2.20 $= 1.87 \times 2.20 \doteq \$4.11$

835.
$$\frac{3\frac{1}{2}}{4\frac{1}{4}}$$

What does $\frac{5}{8}$ mean? It means 5 divided by 8.

So
$$\frac{3\frac{1}{2}}{4\frac{1}{4}} \quad \text{means} \quad 3\frac{1}{2} \div 4\frac{1}{4}$$

At this point it's "old stuff." $\frac{7}{2} \div \frac{17}{4}$

$$\frac{7}{2} \times \frac{4}{17} = \frac{7}{\cancel{2}_1} \times \frac{\cancel{4}^2}{17} = \frac{14}{17}$$

836. Let T = the set of all train cars that Lucy is playing with. Let W = the whole numbers, which is {0, 1, 2, 3, 4, . . .}. Is this a function from T to W? *Assign to each car its weight in grams.*

Suppose the first car weighs 344 grams. The second car weighs 170.3 grams. Oops. This is not a function. To be a function, *each member* of the domain must be assigned an answer in the codomain. But 170.3 is not a whole number. (In symbols, 170.3 $\notin$ W.)

840. Using the results of the previous problem, what would have been Lucy's opportunity cost if she had continued doing the janitorial work, which she could have done for her at $18 per hour?

Opportunity cost is what you give up in order to do something. If she was doing janitorial work, she would be giving up the $2,000 per hour she could be making in doing planning.

Now you can see why brain surgeons don't try to make money mowing other people's lawns.

841. Graph y = 4x – 3 for values of x from 1 to 5.

If x = 1, then y = 4x – 3 becomes y = 4(1) – 3. y = 1	(1, 1)
If x = 2, then y = 4x – 3 becomes y = 4(2) – 3. y = 5	(2, 5)
If x = 3, then y = 4x – 3 becomes y = 4(3) – 3. y = 9	(3, 9)
If x = 4, then y = 4x – 3 becomes y = 4(4) – 3. y = 13	(4, 13)
If x = 5, then y = 4x – 3 becomes y = 4(5) – 3. y = 17	(5, 17)

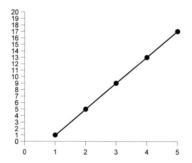

842. In the first hour of Lucky Lucy's Lounge, each order of burritos gave a profit of $21, and each order of chow mein gave a profit of $14. There were 30 more orders for chow mein than for burritos. The total profit in that first hour from these two menu items was $1,365.

How many burrito orders were there?

Let x = the number of burrito orders.
Then x + 30 = the number of chow mein orders.
Then 21x = the total profit from the burrito orders.
Then 14(x + 30) = the total profit from the chow mein orders.
$1,365 was the profit from both items.

$$21x + 14(x + 30) = 1,365$$

and the rest we did in the previous problem. (#741)

850. The wire was very stiff and Lucy's hands were small. It took her 2⅓ minutes to unroll each 10 feet of the wire. How long did it take for her to unroll the 162¼ feet?

We want to convert 162¼ feet into minutes.

$$\frac{162\frac{1}{4} \text{ feet}}{1} \times \frac{2\frac{1}{3} \text{ minutes}}{10 \text{ feet}}$$

$$= 162\frac{1}{4} \times 2\frac{1}{3} \div 10$$

$$= \frac{649}{4} \times \frac{7}{3} \times \frac{1}{10}$$

$$= \frac{4543}{120}$$

$$= 37\frac{103}{120} \text{ minutes}$$

$$\begin{array}{r} 37 \text{ R } 103 \\ 120)\overline{4543} \\ -360 \\ \overline{943} \\ -840 \\ \overline{103} \end{array}$$

851. To the nearest percent how much more is 96 than 3?

96 is 93 more than 3
93 is what percent of 3?
93 = ?% of 3
93 ÷ 3 = 31 = 3100%

Rammy wasn't doing much real estate business because, in part, he set his goals too low. Lucy was teaching him to aim higher.

852. Name a number that stays the same when you square it.

There are two possible answers. $1^2 = 1$ and $0^2 = 0$.

853. In the first month that Luke got his Fred doll, he sent Fred off on 70 adventures. Each month thereafter Luke would send Fred off on 18% more adventures than the previous month. How long would it be before Fred was going on 140 adventures each month?

Translation: How long would it take to double his number of adventures?

Using the Rule of 72: $\frac{72}{18} = 4$

It would take 4 months to double.

900. With 6 diapers she could clean the oil off 8 of her toy engines. How many diapers would it take to clean the oil off of 20 engines?

There are at least two ways to solve this problem.

First way: Use a conversion factor.

$$\frac{20 \text{ engines}}{1} \times \frac{6 \text{ diapers}}{8 \text{ engines}} = \frac{20 \cancel{\text{ engines}}}{1} \times \frac{6 \text{ diapers}}{8 \cancel{\text{ engines}}} = 15 \text{ diapers}$$

Second way: Use a proportion.

$$\frac{6 \text{ diapers}}{8 \text{ engines}} = \frac{x \text{ diapers}}{20 \text{ engines}}$$

Cross multiply $\qquad (6)(20) = 8x$

If you have $\frac{a}{b} = \frac{c}{d}$, then it will be true that $ad = bc$.

Do the arithmetic $\qquad 120 = 8x$
Divide both sides by 8 $\qquad 15 = x$

901. Which of these is correct?
  A) Yesterday I dreamed of pizza.
  B) Yesterday I dreamt of pizza.

Which of these is correct?
  A) Yesterday I kneeled before the King.
  B) Yesterday I knelt before the King.

They are all correct. Some verbs can have more than one past tense. In mathematics 2 + 3 has only one answer.

902. Lucy took Scizzzors outside and put him in the garden. It was Tuesday. She left him there all day. Let the first set S be the set that contains just Scizzzors—{Scizzzors}—and the second set be the set of all numbers R.

Is this a function from S to R? *Assign to each member of S, its temperature on that Tuesday.*

To be a function each member of the domain must be assigned to *exactly one* member of the codomain. Scizzzors was out in the garden all day. In the morning his temperature would be colder than at noon. Scizzzors would be assigned to more than one member of R. It is not a function.

903. Would $252,000 be more or less than a quarter of a million dollars?

A quarter of a million dollars

$\frac{1}{4}$ of $1,000,000

$0.25 \times 1,000,000$

$250,000

Or you could have done it in fractions:

$$\frac{1}{4} \times 1,000,000 = \frac{1}{4} \times \frac{1,000,000}{1} = 250,000$$

Either way, Lucy would be getting a check for over a quarter of a million dollars. Lucy's dad fainted.

Lucy returned all of her dad's tools. She had borrowed them for less than a week.

904. The set S of Lucy's speaking vocabulary is {yucky, locomotive, caboose, toot}. What is the cardinality of this set?

The cardinality of a set is the number of members of that set. The cardinality of {yucky, locomotive, caboose, toot} is 4.

In symbols: card{yucky, locomotive, caboose, toot} = 4 or card S = 4.

The cardinality of Lucy's reading vocabulary is currently around 5,000. Just last week she was reading about building a railroad bridge 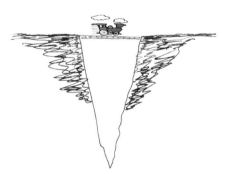 across a chasm. She had to look up the word chasm in the dictionary. It means a gorge or a deep cleft in the earth's surface.

She also learned from the dictionary that it is pronounced KAZ-em. The h in chasm is silent. Many adults don't know this.

905. Lucy puts one-third of her time into reading about railroads. She puts one-fourth of her time into talking with people connected with railroading. She puts one-twelfth of her time working on her train station. Out of 24 hours how much is left for miscellaneous stuff such as eating and sleeping?

$$\frac{1}{3} + \frac{1}{4} + \frac{1}{12} = \frac{4}{12} + \frac{3}{12} + \frac{1}{12} = \frac{8}{12} = \frac{2}{3}$$

If she spends two-thirds of her 24 hours doing this train stuff, she has one-third for the miscellaneous stuff.

$$\frac{1}{3} \times 24 = 8 \text{ hours for eating, sleeping, etc.}$$

In order to save time, Lucy is working on how to eat while she sleeps.

906. 480 is what percent more than 5?

480 is 475 more than 5.

475 is what percent of 5.

475 = ?% of 5

If you don't know both sides of the *of*, then divide the number closest to the *of* into the other number.

$$475 \div 5 = 95 = 9500\%$$

Scizzzors is a lot faster than Lucy at learning things. Lucy, however, has one big advantage. She has a 4-word speaking vocabulary, and this will increase to tens of thousands of words in the next few years. Scizzzors, on the other hand, will never speak a single word.

907. Suppose T (as defined in the previous problem) is the domain and D is the codomain. Suppose the first comic book in D is very special to Luke. He has read it so many times that only it has a torn cover. Is this a function? *Assign Prof. Eldwood's* Locomotive Electrical Diagrams *to the first comic book in D* (the one with the torn cover).

If it is a function then every member of T must be assigned to exactly one member of D.

T is {Prof. Eldwood's *Trains & Today;* Prof. Eldwood's *Choo-choo Trains—A Sociological Perspective;* Prof. Eldwood's *Locomotive Electrical Diagrams;* Prof. Eldwood's *Your Future in Railroading;* Prof. Eldwood's *Why Railroad and Rich Both Start with R;* Prof. Eldwood's *Dining Recipes for Railroad Use*}.

Has every member of T been assigned? No. It is not a function.

910. Name a number that gets smaller when you square it.

There are a zillion possible correct answers!

$(\frac{1}{2})^2 = \frac{1}{4}$

$(\frac{1}{3})^2 = \frac{1}{9}$

$(\frac{2}{5})^2 = \frac{4}{25}$

$(0.1)^2 = 0.01$

Any number between 0 and 1 will work.

In fancy algebra language we will be able to say that $x > x^2$ implies that $0 < x < 1$. But since you haven't had algebra yet, we can't say that.

911. If the receptionist's salary is currently \$18/hour and it is increased by 6% each year, how much will her salary be in one year and in 10 years?
In one year it will be $18 \times 1.06 = \$19.44$

In two years it will be $18 \times 1.06 \times 1.06 = 18 \times (1.06)^2$
In three years, $18 \times (1.06)^3$
. . .
In ten years, $18 \times (1.06)^{10}$,
which if you multiply it out is roughly \$32.24.

912. Let C = {purple, yellow}. Let J = the set of the 31 jockey hats that Fred has. How many possible functions there are from C to J that are not 1-1.

One possible function that is not 1-1 is: *Assign both colors to Fred's first-day-of-the-month hat.*

A second possible function that is not 1-1 is: *Assign both colors to Fred's second-day-of-the-month hat.*

A third possible function that is not 1-1 is: *Assign both colors to Fred's third-day-of-the-month hat.*
. . .
A thirtieth possible function that is not 1-1 is: *Assign both colors to Fred's thirtieth-day-of-the-month hat.*

A thirty-first possible function that is not 1-1 is: *Assign both colors to Fred's last-day-of-the-month hat.*

These 31 functions are the only functions that are not one-to-one.

930. For every value of x that you can name, $y = \frac{x}{4}$ gives you a value of y. For example, if x is 40, then y is 10.

If the domain and codomain are the natural numbers {1, 2, 3, . . .}, is the assignment $y = \frac{x}{4}$ a function?

No. The definition of a function is any rule that assigns to *each* element of the domain exactly one element in the codomain. When x is equal to 1, then y will be $\frac{1}{4}$ which is not an element in the codomain.

931. $3\frac{3}{4} + 4\frac{7}{8} = 3\frac{6}{8} + 4\frac{7}{8} = 7\frac{13}{8} = 7 + 1\frac{5}{8} = 8\frac{5}{8}$

932. Draw a Venn diagram showing two sets: the set of everything that Lucy owns and the set of all the real estate in Kansas.

Lucy owns one piece of real estate—her I Love You Amy Station. She also owns some furniture, some statues, and her clothes.

The two sets overlap since her train station is in both sets.

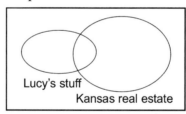

933. Solve
$$5(x + 2) + 12 = 39$$

Distributive law $\qquad$ $5x + 10 + 12 = 39$

Combine like terms $\qquad$ $5x + 22 = 39$

Subtract 22 from both sides $\qquad$ $5x = 17$

Divide both sides by 5 $\qquad$ $x = \frac{17}{5} = 3\frac{2}{5}$

950. What is the square of $6\frac{2}{3}$?

$$(6\frac{2}{3})^2 \;=\; 6\frac{2}{3} \times 6\frac{2}{3} \;=\; \frac{20}{3} \times \frac{20}{3} \;=\; \frac{400}{9} \;=\; 44\frac{4}{9}$$

$$\begin{array}{r} 44 \ \text{R} \ 4 \\ 9\overline{)400} \\ \underline{36} \\ 40 \\ \underline{36} \\ 4 \end{array}$$

951. Fill in the missing entries in this opportunity chart.

	product A	product B
Lucy	$\frac{4}{5}$	$\frac{5}{4}$
receptionist	$\frac{3}{7}$	$\frac{7}{3}$
Luke	1.5	1/1.5 or $\frac{2}{3}$

Index

area and volume
#232. 61
#601. 48
#771. 29

arithmetic
#170. 74
#302. 56
#309. 56
#337. 56
#355. 88
#471. 57
#704. 60
#785. 78
#852. 80
#903. 73
#910. 80

cardinal and ordinal numbers
#132. 90
#571. 52
#775. 18
#904. 18

commutative
#121. 61
#208. 67
#470. 69

conversion factors
#109. 27
#144. 10
#167. 59
#300. 10
#308. 85
#328. 30
#333. 14

#339. 62
#390. 44
#401. 19
#473. 55
#488. 24
#511. 13
#513. 24
#542. 28
#600. 79
#632. 83
#649. 13
#770. 25
#833. 29
#850. 71
#900. 48

discrete vs. continuous
#440. 97
#624. 97

economics laws
#106. 77
#125. 86
#129. 79
#145. 45
#146. 72
#151. 84
#162. 30
#164. 56
#172. 12
#201. 47
#233. 84
#252. 43
#260. 79
#273. 12
#274. 45
#275. 30

Index

#303. 77
#304. 79
#306. 86
#327. 16
#334. 16
#336. 54
#356. 86
#441. 86
#468. 17
#479. 88
#480. 94
#492. 75
#494. 86
#514. 17
#550. 55
#555. 86
#631. 75
#662. 87
#705. 85
#706. 91
#732. 52
#740. 87
#773. 44
#783. 76
#840. 78

English
#118. 19
#331. 46
#338. 28
#491. 57
#515. 57
#556. 58
#646. 58
#654. 58
#901. 58

fractions
#122. 43
#123. 69
#153. 88
#253. 69
#257. 88
#402. 56
#454. 28
#455. 11
#466. 57
#467. 67
#622. 9
#730. 71
#830. 44
#835. 68
#905. 76
#931. 18
#950. 80

functions
#104. 9
#126. 95
#160. 21
#217. 21
#229. 23
#235. 9
#301. 28
#314. 9
#315. 21
#316. 11
#350. 78
#400. 83
#404. 9
#452. 44
#462. 21
#546. 9
#548. 21
#626. 21

#634. 68
#648. 60
#701. 22
#702. 15
#703. 48
#780. 20
#784. 80
#836. 15
#902. 15
#907. 20
#930. 89

graphing
#107. 81
#261. 96
#305. 82
#320. 96
#351. 82
#493. 85
#841. 85

make a guess
#104. 34
#119. 34
#152. 93
#163. 38
#230. 38
#357. 93

percents
#161. 16
#165. 40
#171. 66
#207. 90
#228. 40
#236. 93
#313. 40
#330. 23

#332. 36
#453. 40
#457. 14
#471. 57
#474. 73
#475. 79
#490. 78
#512. 14
#541. 89
#544. 41
#580. 64
#582. 71
#601. 48
#625. 20
#630. 65
#647. 42
#653. 39
#661. 85
#670. 89
#772. 9
#774. 32
#776. 39
#777. 13
#782. 22
#834. 76
#851. 92
#906. 22
#911. 87
#951. 87
#463. 36

proportions
#144. 10
#649. 13